Ecological Studies

Analysis and Synthesis

Edited by

W.D. Billings, Durham (USA) F. Golley, Athens (USA)

O.L. Lange, Würzburg (FRG) J.S. Olson, Oak Ridge (USA)

H. Remmert, Marburg (FRG)

Volume 87

Photo: J.C. Gleize.

Patrick Duncan

Horses and Grasses

The Nutritional Ecology
of Equids and Their
Impact on the Camargue

With 114 Illustrations

Springer-Verlag

New York Berlin Heidelberg London Paris
Tokyo Hong Kong Barcelona Budapest

Patrick Duncan
Director de Recherches
Centre National de la
  Recherche Scientifique
Centre d'Etudes Biologiques de Chize
Villiers-en-Bois, France

This study was carried out at the Station Biologiques de la Tour du Valat.

Poetry excerpts on pp 21, 75, 98, and 129 are from Roy Campbell: Horses on the Camargue. Francisco Campbell Custodio and Ad. Donker (Pty) Ltd. Originally published in 1930 by Faber and Faber, London.

Cover photo: C. Feh.

Library of Congress Cataloging-in-Publication Data
Duncan, Patrick
  Horses and grasses : the nutritional ecology of equids and their impact on the
Camargue / by Patrick Duncan.
    p.   cm. — (Ecological studies ; v. 87)
  Includes bibliographical references and index.
  ISBN 0-387-97543-8
  1. Camargue horse—Food. 2. Camargue horse—Ecology. 3. Wild
horses—France—Camargue—Food. 4. Wild horses—France—Camargue-
-Ecology. 5. Grassland ecology—France—Camargue. 6. Camargue
horse. 7. Grazing—Environmental aspects—France—Camargue.
I. Title. II. Series.
SF293.C28D86   1991
599. 72'5—dc20   91-4190
                                                                     CIP

Printed on acid-free paper

Production managed by Terry Kornak; manufacturing supervised by Jacqui Ashri.

Typeset by Asco Trade Typesetting, Quarry Bay, Hong Kong.
Printed and bound by Edwards Brothers, Inc., Ann Arbor, Michigan.
Printed in the United States of America.

9 8 7 6 5 4 3 2 1

ISBN 0-387-97543-8 Springer-Verlag New York Berlin Heidelberg
ISBN 3-540-97543-8 Springer-Verlag Berlin Heidelberg New York

Melanie Barclay, Marie-Antoinette Diaz, Chantal Heurteaux, Olga Nikonova, and Josiane Xuereb. Jacqueline Crivelli and Georges Vlassis were a great help with some elusive references, and Line Jouventin with the figures. Her professional eye saved some from a fate worse than death—being incomprehensible. Christiane Mauget gave great help with the index.

The text was improved by the comments of Vincent Boy, Professor Bourlière, Tim Clutton-Brock, Alain Crivelli, Rudi Drent, Claudia Feh, Patrick Grillas, Iain Gordon, Dirk Kreulen, Anne-Marie Monard, Bettina von Goldschmidt, and two anonymous referees; the remaining faults are mine.

I am grateful to Claudia Feh, Coco Gleize, Jamie Skinner and M. Jean-Pierre Violet for permission to use their photographs.

The permission of Francisco Campbell to reprint excerpts of Roy Campbell's poetry is deeply appreciated.

Working with the production team at Springer-Verlag was fun: thanks for the considerable help you gave.

My greatest debt is to Alison Duncan who, in addition to computerizing and analyzing some of the data, got the figures into shape, mastered word processing and computerized bibliography programs for the text, and edited the manuscript. She was very observant and patient, though less so than Job, which is perhaps a good thing.

Contents

Appendixes

Section A   Introduction and Background

Chapter 1   Grasslands and Grazing Ungulates

(In the Miocene) . . . grass became common. . . . The horses exploited it and this was surely a crucial factor in their abundance in the later Tertiary and their survival to our times.

George Gaylord Simpson
"Horses"

Grasslands

Steppes, with deserts and savannas, are biomes dominated by the grasses and sedges, Gramineae and Cyperaceae; they cover a large part of the world; see Fig. 1. These grazing lands are of great economic importance because they provide most of the food resources on which domestic and wild ungulates depend.

The grasses appeared in the fossil record in the Cretaceous period, but the most successful families of temperate and tropical grasses became common only in the Tertiary period (Prat 1954). During the dry Miocene epoch, grasslands increased relative to forests (Webb 1977) and now occupy those parts of the lower latitudes which have less than 300 mm of rainfall. With higher rainfall, shrubs and trees become more abundant, and in areas with more than 1,500 mm rainfall, woody vegetation dominates (Nix 1983).

In the dry and humid savannas between the limits 300 and 1,500 mm, there is a balance between herbaceous and woody plants (Noy-Meir 1978). It is well known that fire favors the herbaceous plants and is widely used by indigenous pastoral peoples, as well as by range managers, as a tool to maintain grasslands (Svejcar 1989). Herbivorous animals also play a role (e.g., Kortlandt 1984, McNaughton and Sabuni 1988): the most important species are the "bulldozer" or "mega" herbivores which break and eat

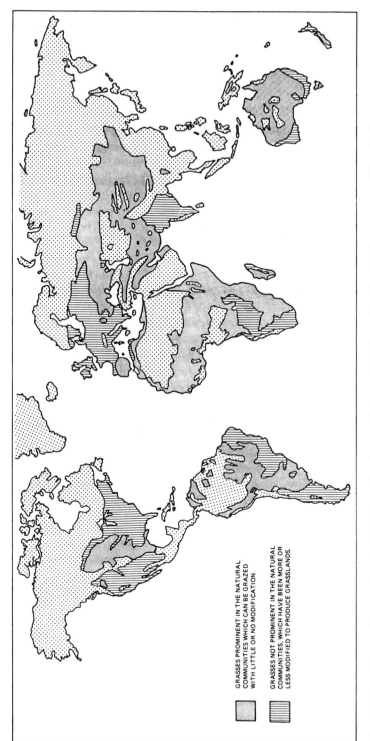

Fig. 1. Map of the main natural and artificial grassland areas of the world. (From Snaydon 1981, courtesy of Elsevier.)

shrubs and trees (e.g., elephants Laws et al. 1975, Owen-Smith 1988), but the smaller browsers and grazers can also play important roles by limiting recruitment of woody plants through the consumption of seedings, especially in ecosystems where these animals are abundant (Good et al. 1990). Grazing by large mammals favors species with effective chemical defenses (Rosenthal and Janzen 1979, Rhoades 1983) and short growth cycles, such as annuals (Risser 1969). Grazing also favors the growth of plants with meristems (growing points) that are at or below the surface of the ground (Harper 1977, p. 435 et seq.), as well as of species and even genotypes with a prostrate growth form (McNaughton 1984).

In addition to influencing both the botanical and physical structure of grasslands and the genetics of the plant populations, ungulates can have important effects on the functioning of grasslands by consuming a large proportion of herbaceous plant production. This happens not only in ecosystems occupied by domestic stock, such as the New Forest (Putman 1986) or the Sahel, but also in near-natural ecosystems, such as the Serengeti, where up to 85% of the standing crop of herbaceous plants in some plant communities is eaten by the ungulates (McNaughton 1979). Primary productivity decreases under very heavy consumption, but it may be stimulated by an intermediate pressure (McNaughton 1983).

Ungulates therefore influence both the dynamics and the functioning of grassland ecosystems. The importance of their impact depends on the animals' densities in relation to the plant resources. Their densities are determined by a wide variety of ecological factors, including predation, weather, etc., but perhaps the key parameter is the value of the vegetation as food for the animals.

The Feeding Ecology of Ungulates: Some Effects of the Plants on the Animals

The value of vegetation as food for an herbivore depends on both the quantity and the quality available. The quantity may vary greatly through the year if plant growth is seasonal. Such variations are very marked both in temperate regions, where the temperature is too low for plant growth in the winters, and in the tropics, where rainfall is too low for growth in the dry seasons. The quantity of herbaceous plants may also vary in space: usually, the water and nutrient status of the soils are the determining factors here. The herb layers of the high ground tend to be short and sparse, while those of the low ground, particularly in wetlands, may be very tall and dense.

Such variations in the abundance of herbaceous plants are of critical importance to animals the size of horses because the variations can limit the rate of food intake (Demment and Greenwood 1988, Ungar and Noy-Meir 1988, Gordon and Illius 1988).

Table 1. Typical Contributions of Cell Contents and Cell
Walls to the Major Plant Tissues

Plant tissue	Cell contents	Cell walls
Fruit	Nearly all	Virtually none
Seed	Mostly	Little
Leaf	Half	Half
Stem	Little	Mostly

The quality of unmanaged swards, too, is heterogeneous in time and space. During growing seasons, the plants are green and lush, but in the non-growing seasons they are mature and fibrous; in marshes, all the aerial parts of many species are dead. However, the concept of quality is a difficult one to define precisely because it varies with the nutrient requirements and the digestive and detoxifying abilities of the animals, as well as with the characteristics of the forages (Van Soest 1982, Chapter 3, Crawley 1983). Large herbivores require nutrients such as protein and digestible carbohydrates from plants: some of these nutrients are located in the "cell contents[1]," which are completely digestible to mammalian herbivores. The concentrations of these nutrients are high in young, green, growing tissues (e.g., Demarquilly and Jarrige 1974) and decrease with the increase in structural fibers such as cellulose, hemicelluloses, and lignin as the plants mature and become stemmy; see Table 1. The structural fibers, which are deposited in the cell walls, are partly digestible by the symbiotic microflora of ungulates, but digestibility is usually low, and for lignin, it is negligible. For these reasons, as plants mature, their quality for herbivores decreases: not only do the concentrations of the nutrients decline, but their digestibilities do too.

In addition to the major nutrients, mentioned above, large ungulates require a wide range of vitamins, minerals, and elements, such as calcium, sodium, and phosphorus. These are generally not limiting in the forages and soils available to free-ranging animals, though special movements (e.g., to salt licks) may sometimes be necessary to obtain adequate amounts of some of these nutrients (Kreulen 1985).

The nutritive value of plants can also be affected by the presence of secondary metabolites. These compounds (e.g., terpenes, tannins) play no apparent role in the normal functioning of the plant cells, but they have an inhibitory effect on feeding by herbivores (Rosenthal and Janzen 1979).

Plant characteristics therefore affect both the daily intake and the availability of the nutrients in forages. The measures of quality most com-

---

[1] The non-fibrous part of plant tissues soluble in neutral detergent (Van Soest 1982, p. 82).

monly used in the agricultural literature are forage digestibility (Blaxter et al. 1961), the protein/fiber ratio and more recently, the product of digestibility and intake (i.e., the rate at which dry matter is extracted from plants, or the digestible dry matter intake; Jarrige 1980, Van Soest 1982). The last measure, while clearly not universally valid, is the most useful general measure of forage quality available.

These spatial variations in the quantity and quality of the herbaceous plants mean that at any one time, herbivores are faced with a wide range of potential foods. If food is ever a limiting resource, then the decisions the animals make as to what they actually eat will have important consequences for their productivity, perhaps for their survival. Empirical evidence shows that grazing animals do tend to feed in swards and on plants with a quality that is better than the average of what is available (e.g., Arnold and Dudzinski 1978, p. 97).

Temporal variations in plant quantity and quality mean that there will be some seasons in which food is abundant and of good quality and others in which food is sparse and/or of poor quality; the phases of the ungulates' reproductive cycle which are nutritionally costly, especially late pregnancy and lactation, are usually timed to coincide with the periods of food abundance. In spite of these adaptations, ungulates are commonly in a negative energy balance during the non-growing seasons. At these times, the animals survive by drawing on their body reserves of fat and protein (e.g., Dauphine 1975, Leader-Williams and Ricketts 1982). Building up their body reserves (or body condition) in the growing seasons is thus a crucial part of their feeding strategies.

These are some of the ways in which plants can affect the behavior (feeding strategies) of the animals which feed on them. Because these effects have consequences for reproduction and survival—the raw material of natural selection—changes in the dominant plants of the planet have had profound effects on the evolution of ungulates (e.g., Guthrie 1984, Van Soest 1982).

Ungulate Digestive Processes

The body sizes of ungulates, used here sensu lato to include the Subungulata, range across three orders of magnitude from 5-kg antelopes to 5,000-kg elephants. These animals feed principally on the vegetative parts of plants—leaves and stems—which are a very abundant food resource, but one which is largely indigestible to mammals because cellulose, the major constituent of leaves and stems, is not attacked by mammalian digestive enzymes.

All ungulates have evolved digestive systems which contain symbiotic microorganisms which can digest cellulose. These digestive systems are of two major types: rumen or cecal (hindgut) fermentation. The largest and

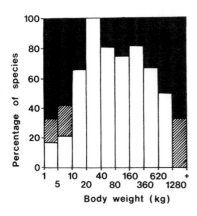

Fig. 2. The distribution of ruminant species (open bars) across the range of ungulate body weights. Ruminants dominate in the range 10–600 kg. The extremes are dominated by hindgut fermenters (black). Hatched bars are non-ruminant foregut fermenters such as hippo. (From Duncan et al. 1990, redrawn from Demment and Van Soest 1985, courtesy of Springer-Verlag.)

the smallest species are mostly hindgut fermenters, but the medium-sized species (10–600 kg) are dominated by the ruminants (Artiodactyla, mainly Bovidae, including both grazing and browsing species); see Fig. 2. The reasons for these asymmetries have been explored in detail by Demment and Van Soest (1985). They argue that the ruminants are more effective at digesting plant tissues in the middle-fiber range, while cecal fermenters are more effcient on low fiber (e.g., the Suidae) and high fiber tissues (elephants, rhinos).

The few medium-sized (20–600 kg), cecal fermenting ungulates include some Suidae, which are omnivorous, and specialized browsing Tapiridae. The grazing species all belong to a single genus, *Equus*, of the order Perissodactyla.

The Evolution of Equids

The Perissodactyla evolved from condylarth stock in the early Cenozoic (see Fig. 3) and radiated into a wide variety of herbivorous ungulates, which dominated the Eocene faunas. Among these were the equids, which were initially browsers. They developed into cursorial grazing ungulates by the Miocene (Janis 1989, Janis et al. in press) when the drier climate of the earth favored the graminoids, grasses, and sedges. These plants have a higher fiber and silica content than the leaves of the browse (trees and bushes) which had dominated the vegetation of the Eocene (Webb 1977, Woodburne and MacFadden 1982). The three-toed grazing equids which evolved in North America, initially exemplified by the genus *Merychippus*, had high-crowned teeth (hypsodonty), an adaptation to feeding on grasses (MacFadden 1976, 1977).

The *Merychippus* line gave rise here to the modern single-toed genus *Equus*, which, by colonizing South America as well as the Old World, achieved a remarkably wide geographical distribution. By the late Pleis-

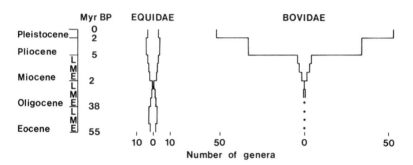

Fig. 3. Generic richness of the equids and bovids in the Cenozoic E—early, M—middle, L—late. (Data from Cifelli 1981.)

tocene, there were five species of *Equus* in North America, one caballine species rather similar to the Przewalski horse, three zebrines, and an onager type (Harris and Porter 1980). There were similar numbers of species in the other continents.

Since the Pleistocene, the once-abundant wild equids have declined, disappearing from the Americas about 8,000 years ago. In the Old World, equids, like many of the large ungulates, declined more slowly.

The reason for their extinction in the Americas, the scene of much of their radiation, has been ascribed to overkilling by people (Martin 1970), to disease, or to climatic change (Martin and Neuner 1978), but it still remains an enigma. A fuller description of the paleoecology of the late Tertiary and Quaternary horses of North America is given by Berger (1986).

In the Old World, the Pleistocene extinctions were fortunately less complete: seven wild equid species have survived into historical times (see Table 2 and Fig. 4), though the ancestor(s) of the domestic horse disappeared along with many other large mammals, including some equids. The Tarpan (the wild horse, *E. ferus*, of Europe, which may have been close to the ancestor of the domestic horse) and the quagga (probably a

Table 2. The Genus *Equus*

E. ferus przewalskii	Przewalski's Mongolian wild horses. The tarpan and domestic horses are very closely related.
E. africanus	African wild asses, domestic donkeys
E. kiang	Kiangs
E. hemionus	Other Asian wild asses
E. burchelli	Plains zebras, quaggas
E. zebra	Mountain zebras
E. grevyi	Grevy's zebras

Based on George and Ryder 1986; Groves 1974.

Fig. 4. Approximate distribution of wild equids in recent times: (1) Przewalski's horse; (2) African wild ass; (3) Asian wild asses, two species; (4) Grevy's zebras; (5) plains zebra; and (6) mountain zebra. (Modified after Groves 1974, Mungall 1979, Short 1975, courtesy of *J. Reprod. Fertility*.)

subspecies of Plains zebra, *E. burchelli*, Hughes 1988) became extinct in the 19th century, and Przewalski's horse, *E. ferus przewalskii*, is now preserved only in zoos.

Of the four species of wild equids in Africa and two in Asia, only two are abundant and widespread, the Kiang (*E. kiang*) and the Plains zebra (*E. burchelli*). The African wild asses and Grevy's zebras are endangered, and the Mountain zebras are vulnerable to extinction.

Today's wild equids are morphologically so similar that even the detailed examination of museum specimens does not allow them to be classified with certainty: in a study of 349 *Equus* skulls on each of which 42 measurements were made, only 90% of individuals were correctly classified into one of the species (Eisenmann and Turlot 1978).

The evolutionary relationships among these species are not yet definitely established (George and Ryder 1986) in spite of the application of modern techniques (e.g., chromosome banding, Ryder and Epel 1978; hemoglobin polymorphisms, Ryder et al. 1979; mitochondrial DNA, George and Ryder 1986). However, these new approaches, coupled with traditional paleontological and morphological studies (Groves 1974), have established that

- Extant equids diverged from a common ancestor 3–5 million years BP
- Zebras are a monophyletic group

- Domestic horses are very close to Przewalski's wild horse; they probably constitute a single species

The main questions which remain concern the Asian wild asses, their place relative to the African wild ass, and the status of the Kiang.

Grazing Bovids and Equids: Competition and Coexistence?

The perissodactyls originated at about the same time as the other major group of medium-sized ungulates, the artiodactyls, about 60 million years ago in the Paleocene (Schaeffer 1948).

Initially dominant, the perissodactyls (particularly the brontotheres and equids) evolved in parallel with the artiodactyls until the Oligocene, when the artiodactyls underwent a great radiation. The perissodactyls declined in generic richness, at least in some continents (Cifelli 1981). Nonetheless, the equids, together with the rhinocerotids were often the commonest ungulates of the Miocene grasslands (Simpson 1951). The recent expansion of the most successful of all artiodactyl families, the Bovidae, was spectacular (see Fig. 3). Many species became adapted to grazing, and the bovids now dominate the medium-sized ungulate fauna of wild and domestic grazing systems of the world, both in species diversity (see Fig. 2; Langer 1987) and in abundance (Cumming 1982). This process has been presented as a classic case of taxonomic, and presumably ecological, displacement: "Exceeded in numbers in early Eocene formations by the perissodactyls, [artiodactyls] have succeeded in far outdistancing their rivals to become the dominant hoofed mammals of later Tertiary and Recent times" (Romer 1966).

Though this interpretation of the paleontological record has been challenged (Cifelli 1981, Janis 1989), it is probable that browsing equids, which became extinct at the end of the Miocene, were outcompeted by browsing ruminant artiodactyls (Janis et al. in press). Further, there is no doubt that the bovid radiation in the Pliocene culminated in their emergence as the dominant herbivores of the grasslands. Nonetheless, grazing equids have persisted into modern times: in the Serengeti ecosystem—one of the few sizable and largely undisturbed ecosystems dominated by grazing mammals—the 200,000 zebra constitute 20% of the biomass of primary consumers and coexist with four highly successful grazing bovids, notably wildebeest, which number over 1 million (Sinclair 1979, Sinclair and Norton-Griffiths 1982).

The grazing bovids and equids share many adaptations to life in grasslands, though these adaptations are analogous rather than homologous, in particular

- Elongated limbs and rather rigid vertebral columns, which allow the animals to move quickly and economically

- Teeth with an increased capacity to grind grass, through hypsodonty, increased complexity of the grinding surface and molarization of the premolars
- Fermentation chambers in the gut to house symbiotic microflora

These facts raise two questions:

1. Why have bovids become the dominant grazers? If they have outcompeted the equids, by what mechanisms have they done so?
2. Why do equids persist?

The evolutionary success of an individual (fitness) depends on the number and quality of offspring raised, relative to intra- or interspecific competitors. The number and quality of offspring will depend on (1) the rate at which the parent can extract nutrients, and (2) the rate of converting the nutrients directly or indirectly into offspring tissues, and (3) the reproductive lifespan of the parent.

Success by these criteria will depend on a large number of physiological and behavioral processes, such as feeding behavior, digestive and reproductive physiology, mating behavior, predator avoidance etc. The present equid–bovid balance is strongly influenced by the responses made by animals in the past to environments different from those in which the survivors live today. The reasons for the dominance of bovids and the persistence of equids can therefore not be known with certainty. Nonetheless, the comparative study of existing species can allow both informed discussions about past interactions and prediction of the results of future ones.

Their economic importance means that the domestic bovids (cattle, sheep, and goats) have been well researched. There is a vast literature on their growth, reproduction, and nutrition in the field of the agricultural sciences. The behavior and ecology of wild bovids have been extensively studied, particularly in Africa (see Kingdon, 1980), where near-natural ecosystems still exist. These two areas of study, agricultural and zoological, are now being drawn together (see the major synthesis of Van Soest 1982), and the biological basis for the evolutionary success and diversity of the ruminants is becoming clearer.

The generally accepted view is that the evolutionary success of the ruminants, and of the bovids in particular, is a consequence of their highly efficient digestive system. "The ascending dominance of ruminants over the non-ruminant herbivores . . . attests to the superiority of fermentation in the stomach over that in the caecum" (Moir 1968). Explanations for the persistence of equids have received much less attention. Though this group of animals has been less researched, the agricultural literature contains the results of thousands of digestion trials using domestic horses; their reproductive system is well studied, and there are many detailed behavioral studies of wild and domestic herds. Two major studies in the late 1960s and early 1970s spanned the agricultural and zoological literature on equids and

Fig. 5. The nutritional model, a qualitative model that has been proposed to describe the differences in daily rates of nutrient extraction (digestible dry matter intake) in equids (dashed line) and in grazing bovids (solid line). (From Duncan et al. 1990, redrawn from Foose 1978, courtesy of Springer-Verlag.)

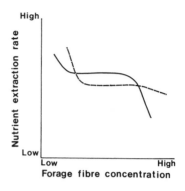

addressed evolutionary and ecological aspects of bovid–equid relations (Bell 1970, 1971; Janis 1976). These two studies set out to explain the persistence of equids. They identified the important contrasts between the digestive physiology of equids and bovids and argued that equids have survived competition with bovids because they can extract nutrients faster than bovids from one type of food, the coarsest forages. These ideas lead to a nutritional model which could explain both the dominance of bovids and the persistence of equids; see Fig. 5.

The diets of zebra and sympatric bovids, such as wildebeest, in the Serengeti (Gwynne and Bell 1968) differed, as predicted by the model. At the nutritionally critical time of year, the dry season, wildebeest specialized on leaves, and zebra on the stems of grasses. This work spawned a number of ideas about interactions between equids and bovids, including the suggestion that equids, by their use of coarse forage, might play a necessary facilitative role for bovids in natural grazing systems (Bell 1971). By implication equid-bovid relations may have been as much mutualistic as competitive over evolutionary time.

Since the development of these ideas, the results of several new studies have been published. These allow the nutritional model and some of these predictions to be tested. Though equids are less efficient digesters of forages, these studies, using both domestic and captive wild animals, show that equids assimilate more nutrients per unit weight and per day than do bovids, right across the range of forage quality, including low-fiber forages with relatively little cell-wall content (see Chapter 3).

The nutritional model therefore accounts well for the fact that equids have survived competition from the grazing ruminants, but it does not account for the numerical ascendancy of the ruminants.

These results show that equids are far from being Miocene relics, which have survived the radiation of the ruminants in the Plio–Pleistocene by feeding on an ecological refuge (the very worst plant tissues, which are unsuitable for the more advanced, complex ruminant digestive system).

They possess a digestive system which is more effective at extracting nutrients from most forages than the ruminant system. Equids must therefore suffer different constraints (e.g., slower reproductive rates, or greater susceptibility to predation etc.), and these should be identified if the ecology of equids, and in particular their interaction with grazing bovids, is to be better understood.

Management Issues

There are currently three main management issues relating to free-ranging equids: (1) conservation of the remaining species, (2) reduction of over-abundant populations of feral animals, and (3) use of domestic and native horses to manage the vegetation of nature reserves.

As described on page 10, four of the surviving equid species are currently endangered or vulnerable to extinction. The reason most often given for their continued decline is competition with domestic stock for essential resources (e.g., water and forage, Wolfe 1979, J. Clutton-Brock 1981, IUCN 1988). Conservation of the remaining 20 or so populations of these species will require increased efforts to manage the animals and their habitats. The captive population of Przewalski's horse has grown from about 50 animals in 1960 to 1,000 in 1990. They are virtually all maintained in the zoos which saved them from extinction, and they must soon be reintroduced into their ancestral range if they are to be preserved as wild, rather than domestic animals. Such conservation and reintroduction operations will succeed only if there is adequate knowledge of the animals' resource requirements, social behavior, and interactions with the plant communities and with other animals, including competitors and predators.

In contrast, feral horses and asses pose real problems of overabundance in every continent, but particularly in the United States (National Research Council [NRC], 1982). After the extinction of the *Equus* species (see page 9), the genus was reintroduced by the Spanish colonizers in the 16th century. Their central role in the history of the colonization of North America, and the resistance by the American Indians has been summarized by Berger (1986) and others cited therein.

Since this time, some feral horse and ass populations have become so abundant (in Australia as well as the Americas) that they are perceived as a threat to wild ruminants and to livestock production systems. These populations are reduced by means which range from the "adopt-a-horse" program to shooting them from helicopters. Feral animals cannot every where be allowed to multiply without check, but modern principles of animal welfare and management demand that such operations be carried out with as little suffering as possible. This, like the conservation of vulnerable populations, again demands an understanding of the animals' ecology and behavior.

these people for a research project involving many different people, often on short stays, was essential.

Professor François Bourlière, Tim Clutton-Brock, and John Krebs helped the study over many years and in many ways; I am very grateful to them. I benefited, too, from discussions with all the coworkers mentioned in the book, as well as with Steve Albon, Joel Berger, Monique Borgerhoff-Mulder, Sylviane Boulot, Tim Caro, Michel Doreau, Joshua Ginsberg, Morris Gosling, Sandy Harcourt, Paul Harvey, Michel-Antoine Leblanc, Thierry Lecomte, William Martin-Rosset, Patricia Moehlman, Ian Nicholson, Marion Petrie, Jacques Poissonnet, Rory Putman, Dan Rubenstein, Y. Ruckebusch, Mike Scott, Kelly Stewart, and P.L. Toutain. Iain Gordon provided stimulating discussion and constructive criticism at a crucial stage.

I am particularly grateful to Coco Gleize, who monitored and managed the herd throughout the period. His knowledge of horses and of the Camargue, his efficiency and dedication to his work allowed many difficult tasks to be carried out with ease. The field data were collected with the unstinting help of many colleagues, as well as Coco Gleize: Melanie Barclay, Peter Elliott, Sandrine Ferrazzini, Laurence Mallinson, Eleanor Mayes, Pamela Moncur, and Jamie Skinner, whose learning curve covered even Provençal, provided valuable assistance over relatively long periods. For permission to use unpublished data, I am grateful to Pamela Cowtan, Chris Gakahu, Pierre Heurteaux, René Lambert, Martin Levertin, Pamela Moncur, and Steve Skelton; and especially to Tom Foose.

The work on horseflies benefited greatly from discussions with and input from Dick Hughes and Jeff Waage; that on social behavior from Anne-Marie Monard, and from Sue Wells and Claudia Feh, who also made major contributions to the collection of data on the use of time and habitats, in addition to their own work. Marc du Lac kindly lent us the Marechal balance for years. I am grateful to him and to the many people from Tour du Valat and from further afield who made it possible to round up a herd of horses gone feral.

Data analysis was initially carried out at the Centre de Recherches en Informatique et Gestion (Université de Montpellier), and benefited from the help of Y. Escouffier and his colleagues. Thereafter, Vincent Boy provided valuable, efficient, and reliable help, as did Steve Albon, with the analysis of growth of the animals and the plants. Kees de Groot provided unbelievably patient tutoring on *WordPerfect*.

Tim Clutton-Brock was kind enough to invite me to spend two sabbaticals at the Large Animal Research Group at Cambridge. These stays provided a wealth of ideas in an ideal atmosphere for writing-up. I also am grateful to the President of Wolfson College for a Visiting Fellowship, to Professor Gabriel Horn (Department of Zoology), and to Dr. Paul Howell.

If the fieldwork was usually fun and the analysis interesting, writing this book was a toil which would never have been finished without the help of

Acknowledgments

Working on the horses in the Camargue was a formative and exciting experience for a young zoologist, especially when based at the Station Biologique de la Tour du Valat with a team of people who possess a remarkable understanding of the ecology of the delta. They and many people outside the station, by sharing their knowledge and skills and giving of their time, have made this project possible. Space does not allow each to be mentioned individually here, but their contributions are no less appreciated.

The people of the Camargue, who have conserved the horses and the marshes, taught me about their animals when, thanks initially to Jean-Claude "Coco" Gleize, I rode with them to fetch their herds of cattle from the marshes. They showed me what it means to live integrated management in a European wetland, the difficult double act of conservation and development. Jacques Mailhan, in particular, has always been interested in our approach to the ecological role of the horses and cattle and helped in many ways.

Bettina von Goldschmidt and Professor B. Tschanz had the idea of this study; Dr. Luc Hoffmann put it into effect with the technical assistance of René Lambert and broadened the original concept to cover the ecological impact of horses on the wetlands. The study was funded by the Fondation Tour du Valat, and administered successively by Michel Pont and Jean Paul Taris. Pierre Jouventin and the Centre National de la Recherche Scientifique provided support in the final stages. The patience and understanding of all these, Przewalski's horse, which is closely related to the ancestor of the domestic horse, is extinct in the wild.

The situation of domestic horses is very different. In spite of the invention of the internal combustion engine, there are still millions of them. They are used for transport and traction in the less-developed parts of the world, and for leisure activities in developed countries. In every continent, domestic horses have escaped and established feral populations. Where low human densities have meant that these feral populations have been left uncontrolled for some time, as in North America and Australia, they have become so abundant as to be perceived as pests by some people.

Equids therefore pose two contrasting management problems: how to conserve and increase the sizes of the populations of threatened wild species, and how to reduce the feral populations efficiently and humanely so that they are in balance with the, often artificial, ecosystems in which they live. To succeed, management must be based on a thorough understanding of the equids, their feeding and social behavior, and on the ways they interact with the plant communities they live in and from.

The ecology and behavior of ruminants and their interactions with plant communities have been thoroughly studied for economic reasons. Equids are much less well known. In the Camargue, France, a herd of horses was experimentally released into a nature reserve and allowed to increase without human intervention. This book brings together the results of this long-term study. It is written as a contribution to our understanding of the interactions between grasslands and one of the herbivores which shaped and shapes them. I hope that it will prove useful not only to specialists in this field, but also to the community of people working to save the wild species and to manage the feral populations of equids.

Preface

Grasslands constitute one of the planet's major biomes (Chapter 1, Fig. 1) and have done so for 30 million years (Webb 1977). The plants of grasslands have coevolved with grazing animals during this period, and this interaction has shaped the growth-forms, chemical composition, and patterns of reproduction of the plants (Crawley 1983). In the Pleistocene and Recent periods, the communities of grazing ungulates have been dominated by ruminants such as the Bovidae (antelopes, cattle, bisons,[a] etc.), both in wild and domestic grazing systems.

The ruminants are a highly successful artiodactyl group which evolved recently; before the Pliocene, the commonest ungulates of the grasslands, at least in some continents, belonged to the order Perissodactyla, rhinocerotids and, particularly, equids (Simpson 1951, Cifelli 1981). The equids remained numerically abundant, if species-poor, up to the end of the Pleistocene, especially in the temperate grasslands of North America.

Today there are seven species of equids, all belonging to one genus (*Equus*; see Chapter 1, Table 2), compared with 49 genera of Bovidae (Walker 1968). The only large populations of wild equids are the Plains zebras of East Africa, and kiangs of Mongolia and north-western China. Five of the seven species are vulnerable or threatened with extinction, and one of

[a]Latin names are in Appendix 1.

vii

For Alison Duncan
who cares for horses, grasses, and the Camargue

Many protected areas are composed partly or wholly of grasslands. The artificialization of these ecosystems due to the extinction of certain wild ungulates, and to the restriction of movements of surviving populations can cause problems which lead the managers of the areas to intervene. In Southern Africa, elephant populations which are perceived to be over-abundant are commonly culled (Jewell and Holt 1981). In European nature reserves, the absence of large ungulates leads to plant successional processes which are incompatible with the aims of some of the reserves. Domestic ungulates, particularly cattle and horses, have been introduced to replace the extinct wild ones (Gordon and Duncan 1988, Gordon et al. 1990). The success of such interventions will naturally depend, at least in part, on a proper understanding of the processes involved in interactions between plants and their ungulate consumers.

Success in the conservation of rare species and in the management of both overabundant populations and free-ranging populations of herbivores in nature reserves requires a thorough knowledge of the animals' ecology and behavior.

Previous Studies of Equid Ecology and Behavior

There are therefore strong applied as well as fundamental reasons for the study of the ecology and behavior of equids. Previous studies fall into two groups: (1) investigations of wild or feral populations, and (2) studies of domestic animals. The former have concentrated on behavior (e.g. Klingel 1972, 1977; Rubenstein 1981; Berger 1986; Ginsberg 1988 and have shown that in arid areas, where densities are low, males generally defend resource territories which contain food or water supplies. Most groups of animals are stable for only a few days at a time; none last for more than a few months. Classic examples of this system are Grevy's zebras (Ginsberg 1988) and feral asses in the deserts of the United States (Woodward 1979). In the species of more mesic habitats, such as Plains and Mountain zebras and feral horses, the typical social system is one based on closed mem-bership groups of females, often fewer than three, with their recent offspring. A single stallion has mating rights over these females until he is replaced by another male (Klingel 1967, 1968, 1969; Berger 1986). In ex-ceptional circumstances, a group may have two or more adult stallions (Miller and Denniston 1979); territorial defense is rare (Rubenstein 1981); home ranges may overlap completely; and the bands may associate tem-porarily (Klingel 1967) or permanently (Ginsberg 1988). Assemblages of thousands of Plains zebras can form, as on the Serengeti Plains (Klingel 1972).

Recent studies of the behavior of equids have emphasized ecological factors as determinants of the distribution patterns of females, and therefore of the social organization of a population (Rubenstein 1986). However, there have been relatively few studies of the basic ecology of equids. The habitats used by equids have been described for several populations (e.g. Salter and Hudson 1978, Rubenstein 1981, Berger 1986, Keiper 1980, Gakahu 1982). Equids live at varying densities in habitats as different as deserts and wetlands. The causes of these variations in density, which are presumably related ultimately to the food supply, have received little attention (but see Duncan 1983, Gordon 1989a,b,c).

In North America, where mustangs potentially compete with domestic horses and cattle, there have been a number of studies of equid diets. Horses generally feed on graminoids; when these are not abundant, they switch to dicotyledons—trees and shrubs, as well as forbs (e.g., Vavra and Sneva 1978). African equids are also graminoid specialists (Gwynne and Bell 1968, Owaga 1975). Neither the causes of the animals' preference for graminoids nor the nutritional consequences of a switch to forbs have been determined. There have been few comparative studies of the feeding ecology of wild equids and grazing bovids (Gwynne and Bell 1969, Owaga 1975).

Detailed studies have been made of the population ecology of both wild equids such as the Plains zebra of the Kruger National Park in southern Africa (Smuts 1976a) and feral horses such as on Sable Island, off the east coast of Canada (Welsh 1975). The demographic data from a number of studies have been collated and form the basis for a theoretical approach to mustang demography (Eberhardt et al. 1982), which addresses the causes of variations in recruitment and population growth rates. Bovid populations, at least sometimes, are limited by their food supply (Sinclair et al. 1985), but very little is known of the factors limiting equid populations. Sinclair and Norton-Griffiths (1982) have suggested that the largest zebra population, the Plains zebra of the Serengeti, is uncoupled from its food supply and is limited by other factors such as predation or social constraints. Berger (1986) has shown that social competition reduces the recruitment rate of feral horses in North America, though probably not their population size.

Wild equids are not ideal subjects for research on behavior and ecology: they live in remote places, in large populations, and are often difficult to approach. Much of the research on equids has therefore been done on domestic horses: for these, there is a wealth of quantitative information on genetics and breeding, reproductive and digestive physiology, diseases and parasitism, (see Evans et al. 1977). Two recent comparative studies have examined the feeding behavior and habitat use of sympatric horses and cattle (Putman 1986, Gordon 1989a,b,c). There has also been a large number of studies of the social behavior of individuals (e.g. Tyler 1972, von Goldschmidt-Rothschild and Tschanz 1978, Houpt and Keiper 1982) and of their feeding behavior (Carson and Wood-Gush 1983, Hawkes et al. 1985). These studies allow experimental work because the horses and their environment can be manipulated easily, but their results may be difficult to interpret because the conditions are unnatural.

Long-term quantitative data on the behavior and ecology of individually known horses which suffer little or no human interference can clarify the results of both these sets of studies of wild and domestic equids. Such studies require populations which are relatively small, accessible, and approachable so that detailed long-term information can be obtained on the life histories of individuals; investigations of this kind have been carried out on feral horses (Berger 1986, Rubenstein 1986), on feral asses (Moehlman 1974), and on Grevy's zebras (Ginsberg 1988). This approach has provided important insights into the behavior and ecology of equids.

This Study

The work reported here has been done on a herd of Camargue horses, which live in the delta of the Rhône River in southern France, one of the foremost conservation areas of Europe. The Camargue is best known for its large and diverse waterbird populations. The horses belong to a hardy local breed, which has grazed the Camargue for millennia.

The study began in the early 1970s, before any published information was available on the behavior of unmanaged horses. The aim was to describe in detail the development of a natural social system in a Camargue horse herd when traditional management ceased. It started with a small herd (14) on a large area (300 hectares—ha), in order to have 5 years without serious food shortages. In addition to behavioral studies, research on the interactions between the horses and their environment was initiated, both to provide information on the environmental constraints under which the animals operated, and to describe the impact of the animals on the range. This impact determines the uses to which the horses can be put for the conservation of the birds and the ecosystems in which they live in the Camargue and elsewhere.

The herd increased rapidly (see Fig. 6), and density rose by a factor of five over 7 years. In 1981, there were nearly a hundred animals at a density of over 0.25 horses/ha. They had reached the food ceiling, for the *annual production* (the change in weight of the herd, plus the weight of individuals born, exported, or dead), which had increased until 1979, decreased to close to zero in 1982. The population therefore went through an eruptive cycle of the type that has been described for many populations of ungulates introduced into a new habitat (e.g., Caughley 1970a). The animals' feeding behavior, nutrition, growth, reproduction, social organization, and impact on the habitat were analyzed by a number of research workers during this population eruption. The circumstances of this study allowed precise records to be made of this interaction; these records are difficult or impossible to make on wild populations.

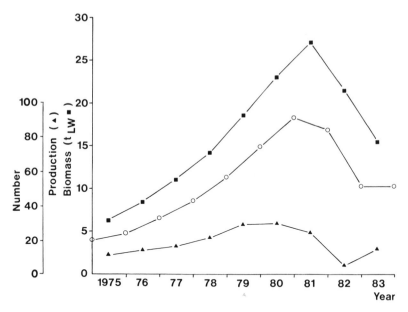

Fig. 6. The average biomass (filled squares) and total production (filled triangles) of the herd in each ecological year (1.9–31.8). The number of horses on the first of September of each year is also given (open circles).

The Structure and Contents of This Book

In this book, I analyze the factors affecting food intake and digestibility in equids. I then bring together the results of this long-term study and give an account, schematized in Table 3, of the relationships between the horses and their habitat during the population eruption. The results are compared with what is known of free-ranging equids elsewhere, with emphasis on the harem-forming species, such as Plains and Mountain zebras and feral horses. Particular attention is paid to contrasts and convergences between equids and the grazing bovids.

After the introductory Section A, strategies for harvesting nutrients are examined in Section B. The processes of digestion and assimilation, which ultimately control feeding behavior and production, are reviewed in Chapter 3 (the first in Section B). For free-ranging animals, high rates of nutrient extraction depend on the animals' ability to harvest large quantities of high-quality food. Chapter 4 examines individual and seasonal variations in diets, food selection, and food quality, and it evaluates the horses' ability to achieve high rates of food intake on the range. The ranging and feeding behavior of equids is described in Chapter 5, which covers the selection of feeding habitat, the allocation of time to feeding, and the temporal organization of feeding behavior.

Table 3. Interrelated Aspects of the Nutritional Ecology of Equids; and the Chapters in Which they are Considered.

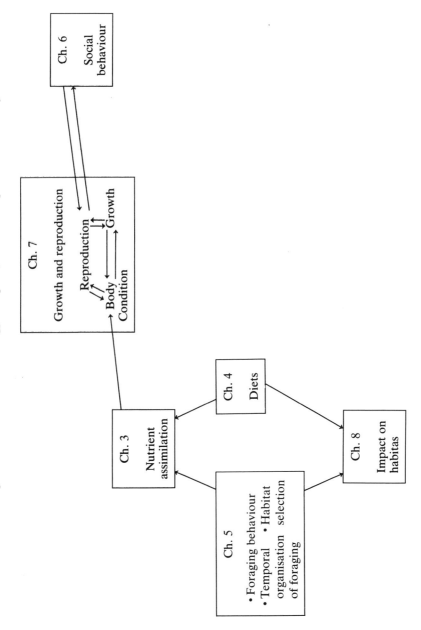

The uses to which the nutrients are put are examined in Section C. The social system is described in Chapter 6, and its dynamics are analyzed in terms of the breeding strategies of males and females. In Chapter 7, rates of reproduction and growth in liveweight are described and compared, both among animals of different social status, and between periods of high and low food availability.

In the final section (D), the role of equids in grazing systems is examined. In Chapter 8, it is shown that they can play a major role in structuring plant communities in wetland and dryland habitats. In the final chapter (9), the factors limiting equid populations, and the evidence for competition and facilitation between equids and other grazers are reviewed. Some conclusions of this research are drawn, for both the conservation of the surviving populations of wild species and the management of potentially overabundant populations of feral equids.

Chapter 2 The Horses and the Camargue

Horses on the Camargue
In the grey wastes of dread,
The haunt of shattered gulls where nothing moves
But in a shroud of silence like the dead,
I heard a sudden harmony of hooves,
And, turning, saw afar
A hundred snowy horses unconfined,
The silver runaways of Neptune's ear
Racing, spray-curled, like waves before the wind.
Sons of the Mistral, fleet.

Roy Campbell

The Horses

The Camargue is a medium-sized horse (1.35–1.45 m at the shoulder, 350–500 kg). It is remarkable for its coat color (uniformly white in adults), its heavy limb bones, and its broad hooves, which are apparently adaptations to the wetlands in which it lives. The historical record suggests that the morphology and color of the Camargue horses today (see photographic plates) are little different from what they were hundreds and perhaps thousands of years ago (see Appendix 2). This stability implies that modern husbandry practices have had little effect on the animals, in spite of the intensive crossbreeding practiced in the l9th century to produce horses for the light cavalry. By 1950, most of the introduced characteristics had disappeared, and in 1978, when the breed was officially recognized by the Haras Nationaux (Bideault 1978), the animals were morphologically closely similar to Poulle's description in 1817, before the crossbreeding began. However the results of blood typing (Kaminski and Duncan 1981) show that the Camargue horses have not the high degree of genetic heterozygosity characteristic of hardy breeds of European ponies such as the New Forest. In this respect, they are intermediate between the ponies and highly selected breeds, such as the Arabians and Thoroughbreds.

Camargue horses have always ranged extensively, with minimal husbandry. They can therefore be expected to have maintained adaptive responses to the environmental pressures they live under, in particular to the seasonally sparse and poor forage, to the weather, and to attacks by the biting insects (cf. Rozin and Kalat, 1971). The punitive work (threshing) required by their owners in the past,[a] and the type of low-input husbandry they employed must also have contributed to the breed's hardiness. More recently, there have been both informal and official efforts to select for the traditional characteristics, in particular the uniform white coat color, which is unusual in breeds of horses.

Today, there are about 3,500 horses in the Camargue: they are kept by their owners in small herds of about 15 mares with their recent offspring. There is only one herd with more than 100 animals (on the Giraud estate).

In the 10 years before this study began, the Tour du Valat herd had been more closely managed than previously. Between ten and twenty mares were maintained, with the aim of producing foals for sale, but the mares were not handled or given veterinary treatment. Supplementary food was limited to occasional rations of coarse hay in particularly harsh weather.

During the study, the herd lived on unimproved vegetation. The animals were not managed, so their feeding and reproduction were natural. During the initial 5 years, which had high rainfall, the population reached 57 horses (see Chapter 1, Fig. 6), and the impact of the horses was slight over most of the range. The robust response of the plants was not surprising, for they had a long history of grazing by sheep, goats, and horses.

At this stage, the horses had developed a social system similar to that of Plains zebras, and a great deal of information had been collected about their behavior and ecology. However, the low density of the horses relative to their food resources meant that the relevance of the results to naturally regulated populations was doubtful. It was therefore decided to allow the herd to continue its increase and to use the unique opportunity provided by

---

[a] Teams of six pairs of mares (rodos de rosso) working for 10 hours a day covered approximately 80 km in the threshing ring. The season lasted about 2 months (Naudot 1977).

its accessibility to obtain detailed data on the animals' diets, condition, social behavior, liveweight growth and reproduction, and impact on the range.

It would clearly have been unethical to allow animals in this enclosed herd, which had no predators, to die of starvation, so two pastures of 80 ha each were held in reserve for animals that became dangerously thin. The constraints were that (1) heavy grazing should not cause a long term reduction in the productivity of the rangelands, and (2) horses should not die of starvation.

It was expected that the food shortages would (1) affect the females most, (2) reduce the rate of recruitment, and (3) result in increasingly intense reproductive competition between the males, which would (4) increase their energetic costs to something approaching the costs of lactation to females. Management of the horses required the removal of individuals which became very thin (condition class 6; Chapter 7, Table 2). This was generally done by rounding up the whole herd with riding horses. These round-ups were also used in order to measure body weight and to take blood samples for blood typing, with the aim of determining the paternity of the foals, an exercise which was made possible by collaboration with the Equine Research Station at Newmarket, United Kingdom, and the Centre National de la Recherche Scientifique at Gif-sur-Yvette, France.

The population rapidly reached the food ceiling (cf. Chapter 1, Fig. 6), and by 1982, there was enough food only for maintenance of the herd. To give warning of overgrazing, peak plant biomass was measured in plots protected from grazing; and plant cover was measured on the grazed range. The heavy grazing pressure caused a sharp decline in the plant cover of the grasslands (see Chapter 8): in order to avoid long-term damage to the range, the number of horses was reduced to half of the peak number in September 1982, and the herd was moved onto reserve pastures in the growing seasons of the following years. In spite of this careful management, the animals remained in poor condition in 1983; without such management, there would certainly have been a major die-off.

The Camargue and the Study Area

This section provides both a brief description of the Camargue region and information on physical and biological parameters that are of particular importance to the horses. A more detailed account can be found in Hoffmann (1958), Picon (1978), Lemaire et al. (1987), Boulot (in press) and other references in Biber (1975).

After a preliminary section on the geology and topography of the area, its history and current management, the soils, landscape, climate, water depths, and vegetation are described. The trophic impact of the more important vertebrate herbivores is estimated.

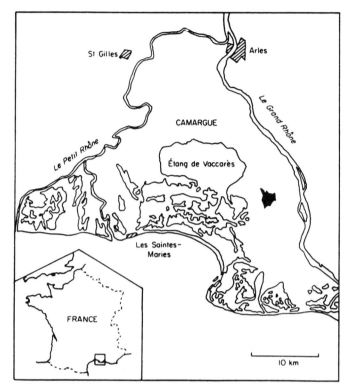

Fig. 1. Geographical location of the Camargue; The Tour du Valat estate is in black. (From Bassett 1978, courtesy of Blackwell Scientific Publications.)

Since the disappearance of wolves in about 1850, the only predators which exploit the horses are people and the parasitic nematodes, ticks, leeches, and *Diptera*—especially midges, mosquitoes, and horseflies. The phenology and (where possible) the abundance of these invertebrates are described.

The Camargue is the 145,000 ha delta of the Rhône River in Mediterranean southern France; see Fig. 1. It was formed from alluvial matter deposited by the Rhône during the Flandrian stage of the Upper Pleistocene, and its formation continued during the Recent (Holocene) period. Most of the delta is therefore relatively young (<20,000 years old, Duboul-Razavet 1955).

The flat topography (the slope from Arles to the sea is only a few centimeters per kilometer) has meant that the course of the Rhône has been unstable, and at least six major courses different from the present one have been detected south of Arles on aerial photographs of the delta (Kruit 1955). The more recent ones are still dominant features of the landscape (e.g., the Bras de Fer, the Rhône d'Ulmet). The coastline, too, has been unstable during the recent past: it cut across the center of the present delta at the level of the Bois des Rieges during the Flandrian (Duboul-Razavet 1955), and the coastline at Beauduc continues to extend west, while that at les Saintes-Maires-de-la-Mer (see Fig. 1) has moved several kilometers northward since the Middle Ages. However this instability, typical of the deltas of many great rivers, has been tamed: since 1870, dikes have maintained the two arms of the lower Rhone in their present courses, and a third dike has prevented further incursions by the sea.

These dikes have artificialized the hydrology of the delta, thus making possible the development of intensive arable agriculture in large areas of the Camargue through the extension of interlocking systems of drainage and irrigation canals.

Within the delta, there is little relief: the lowest point is the sump of the Vaccarès (-2 m above mean sea level, a.m.s.l.), and the highest points occur on the littoral dunes, which may reach 4 m a.m.s.l. Much of the Camargue varies in altitude by only 1–2 m, but small differences in elevation can cause great changes in the water regime, the vegetation, and the land-use practices.

In the south, industrial salinas cover some 21,000 ha. They not only produce a very large salt harvest (over 1 million tons/year) but also provide feeding, roosting, and nesting habitats for a large number and variety of birds, including flamingos (*Phoenicopterus ruber*), shelduck (*Tadorna tadorna*), and many species of waders.

Arable agriculture occupies a further 52,000 ha; industrial and urban areas cover 8,000 ha; and the rest, some 64,000 ha, is unimproved land, mostly beaches, marshes, and salt flats, which are used for tourism, nature protection, hunting, fishing, and the grazing of local breeds of horses and cattle (see photographic plates). The agricultural and industrial development of the Camargue has meant that the grazing lands have declined considerably in recent times (by 34% between 1942 and 1984, per Lemaire et al. 1987) and today the herds live in enclosures bounded by fences and canals: few are over 500 ha in extent.

Land use in the central 85,000 ha of the delta is under the surveillance of the Parc Naturel Régional de Camargue, the object of which is to ensure that the area is developed economically in ways which allow the conservation of the natural and cultural resources of the area. Parts of the delta have been declared a Biosphere Reserve, a Ramsar Convention, and a European Diploma site.

Land is owned by both the private and the public sectors. The largest holding in public ownership is the Réserve Nationale de Camargue, 13,000 ha, which is managed by the Société Nationale de Protection de la Nature for the protection of nature and for scientific research. The total area of "réserves", with management policies that range from nature protection to regular hunting, is 21,000 ha, 14% of the delta.

The major part of the Camargue is privately owned: the salt pans belong to the Compagnie des Salins de Midi et de l'Est, and the agricultural areas, together with the grazing lands outside the reserves, belong to private land-owners.

One of the large areas of undeveloped land is the Tour du Valat-Petit Badon (1,850 ha), which belongs to the Fondation Sansouire, established in 1978 by Luc Hoffmann. The research and management carried out at Tour du Valat are funded principally by the sister Fondation Tour du Valat, the aim of which is to "conduct and promote scientific research with special reference to conservation and management of wetland communities in the Camargue and around the Mediterranean." The studies reported here were carried out in the eastern part of this estate, on some 300 ha, which contained representative areas of most of the major types of habitat which occur in this part of the Camargue, as well as a small area of fallow rice fields.

Soils of the Camargue

Of alluvial origin, the soils consist mainly of silts composed of a variable mixture of clay, mainly illite and kaolinite, very fine crystals of quartz, and calcium carbonate. The soils are remarkable for their low organic matter content; this is usually less than 1% and A_0 horizons are found only on the highest ground and in marshes (Duboul-Razavet 1955; Heurteaux 1969). The other important feature of these soils is their salt content, which is so high in the central and southern Camargue that the soils are classed as solonchaks (Kubiena 1953, Bassett 1978).

Bassett (1978) has made a thorough study of the soils of the eastern half of Tour du Valat as part of an investigation of soil/plant relationships. These soils are basic, and pH values average 7.7–8.4 in different vegetation types. Nitrogen concentrations are low (ranging between 0.12 and 0.71%); the highest values are obtained in the soils of the marshes where some organic matter does accumulate. There are high levels of total exchange-able bases: these average between 1.7 and 2.4 g per 100 g soil in the different vegetation types. The greater part of the exchange capacity is made up of calcium ions (1.4–1.6/100 g soil), though sodium ions are locally abun-dant (0.035–1.0/100 g soil).

Classification and Mapping of the Study Area

A landscape classification of the Tour du Valat was made by P.M. Rogers (1981 and personal communication, 1977). This approach is hierarchical

Fig. 2. Landscape classification of the study area. For descriptions of facets (5, 4 etc.), see Table 1 (+315—numbered stake for vegetation sampling; A—modified by agriculture; B—scrub and woodland.) (From Rogers 1981, courtesy of Black-well Scientific Publications.)

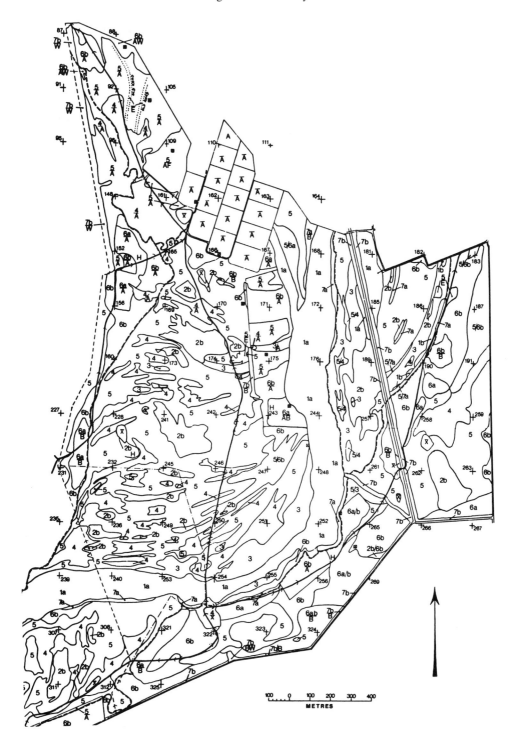

and consists of differentiating Land Systems and the Land Facets which make up the systems, based on air photos, using a wide range of natural features. These features include land form and topography; soils and their hydrological characteristics; vegetation pattern, structure, and cover; and land use (Mabutt 1968).

The study area is located in the eastern half of Tour du Valat and covered 298 ha, which increased to 335 ha in March 1976. The horses had free access to the whole area, which was enclosed by fences and canals. The landscape map of the study area is shown in Fig. 2, and the facets are described in Table 1.

Table 1. The Land Facets, with Descriptions and Areas

Facet	Description	Area[b] (ha)	Coverage (%)
1	*Deep marsh*: dominated by the reeds *Phragmites* and *Scirpus* (Groups E/F and G)	26.1	(7.8)
2	*Shallow marsh*: *Tamarix* with herb layer dominated by *Aeluropus* and *Juncus gerardi* (Group B)	24.6	(7.3)
F	*Fallow fields*: heterogeneous, ruderal vegetation, dominated by *Aster, Paspalum, Poa*, and *Trifolium* (Group H)	20.5	(6.1)
3	*Wet enganes*: dominated by *Arthrocnemum*, often with *Juncus subulatus* (Group D)	21.1	(6.3)
4	*Sansouïre*[a]: bare ground with *Arthrocnemum glaucum*; other plants absent or very rare (Group C)	35.6	(10.6)
5	*Enganes*[a]: dominated by *Arthrocnemum*, with *Halimione, Puccinellia* and a lot of bare ground (Group D)	123.5	(36.8)
6b (6 in text)	*Halophyte grassland*: species-rich sward with many annual grasses (*Bromus, Hordeum*) perennial halophytes (*Limonium, Halimione*) (Group A_2)	51.7	(15.4)
6a, 7a, 7b (7 in text)	*Coarse grassland*: dominated by *Brachypodium* and *Phillyrea* (Group A_1)	31.5	(9.4)
	Roads and excavations	0.97	(0.3)
	Total	335.6	100.0

Note: The vegetation group most similar to the facet is listed at the end of the description (Rogers 1979, 1981).
[a] Local names for these types of habitat, which occur on salt flats.
[b] Where an area was identified as two facets (e.g., 4/5), half was allocated to each facet.

The range of land types found in the eastern Camargue is represented here: from the high ground with coarse grasses and bushes, through more open, saline habitats, to marshes, which can be a meter deep. Although the horses were enclosed, they had the opportunity to choose among all the habitats that horses could have used before the Camargue was fenced (i.e., except the riverine forest and the maritime dunes).

Climate of the Camargue

The Camargue, like the rest of the Mediterranean region, has cool, wet winters and warm, dry summers. Detailed information on the climate is given in Appendix 3.

Winter temperatures fall below 0°C on an average of 5 days a month, December to February. Snow fell twice in 10 years, but the mistral wind, commonly >100 km/hr, compounds the animals' heat losses.

Rainfall is extremely variable: in the decade considered in this book, it varied from 405 to 950 mm. Annual rainfall totals have not been calculated for the calendar years of the study, but rather for the periods from September to August—the "ecological" year; this has been done because the rain falling in autumn influences plant production of the following growing season (spring and summer) more than that of the calendar year in which it falls (cf. Corre 1975). The totals are given in Table 2, together with the average annual rainfall for the years 1944–1973.

The climatic index used in the region is Emberger's Q_2 (Emberger 1955). The values of Q_2 for each ecological year are presented with the values for the previous 16 years, given in Berger et al. (1978); see Fig. 3.

Table 2. Mean annual rainfall 1944–1973 and annual rainfall (Sept–Aug) for the years 1974–1983

Year	Rainfall (mm)
1944–73	582
1974	936
1975	506
1976	628
1977	950
1978	816
1979	437
1980	777
1981	463
1982	405
1983	588

Source: Heurteaux 1975 and Tour du Valat Met. Records).

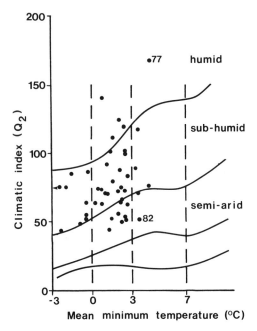

Fig. 3. Climatic indexes for Tour du Valat, for the years 1944–1986. Q_2 is Emberger's climatic index; see Appendix 3; mean minimum temperature refers to the coldest month. (Redrawn from Berger et al. 1978, courtesy of Revue d'Ecologie—Terre et Vie.)

The climate is usually subhumid, but varies from the humid to the semi-arid. The year 1977 was exceptionally humid, while 1982 was the most arid of the whole set.

In conclusion, the climate of the Tour du Valat is generally Mediterranean subhumid, with cool winters. There are great variations in climate from year to year; this is particularly true of the rainfall. Variability is greatest in the dry summer months, which could exaggerate year-to-year differences in plant production. The first 5 years tended to be more humid than the average, the last 5 more arid. Winter temperatures are never very low, and snow is rare, but high winds increase the heat losses of large mammals. In the summer, air temperatures are below the body temperature of large mammals (37°C in horses); it is unlikely that thermoregulation is a major constraint in this season because water is abundant.

Water Levels

With heavy rain, the silty soils become waterlogged, and surface water collects in Facets 1 and 2, the Deep and Shallow marshes. In exceptional circumstances, after very heavy rains such as those that occurred in Octo-

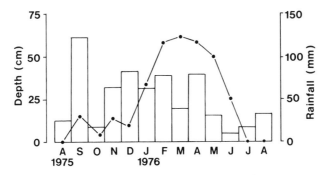

Fig. 4. Maximum depth of the Relongues marsh and rainfall, 1975–1976.

ber 1976, Facets 3, 4, and 5 also became flooded. Under these conditions, about 75% of the area was under water.

Seasonal changes in water depth, measured at the deepest part of the Deep Marsh, are shown in Fig. 4. There were more than 50 cm of water at some stage in each year, and the duration of flooding ranged between 160 and 345 days. This hydrological regime allowed the development of the dense reedbeds in the low-lying parts of the study area.

Vegetation

The plants of the Camargue are well-known from the work of Molinier and Tallon. They have described the phytosociological associations using the Braun-Blanquet (1952) system, listed the vascular plants, and provided a map of the distribution of the associations (Molinier and Tallon 1970, 1974 and Molinier and Devaux 1978). About 950 species and subspecies have been recorded, most of them characteristic of Mediterranean wetland or saline habitats; an important exception is the set which occur on the highest, best drained ground (e.g., *Juniperus*, *Brachypodium*, *Phillyrea*), which are xerophytic, Garrigue plants (Molinier and Tallon 1970).

Floristic Classification of the Vegetation

The plant associations, and plant–soil relationships in the study area were described using multivariate methods (Bassett 1978). Vegetation and soils were sampled at randomly chosen points in the study area. The vegetation was classified by indicator species analysis (Hill et al. 1975), and the dendogram shown in Fig. 5 was obtained at the eight-group level. One group (E) contained only two stands and, for the purposes of this study of horses, was considered together with the floristically similar Group F, while Group A, the largest, was subdivided into A_1 and A_2, the groups obtained at the 16-group level. Plant species showing significant association with the eight groups are shown in Table 3.

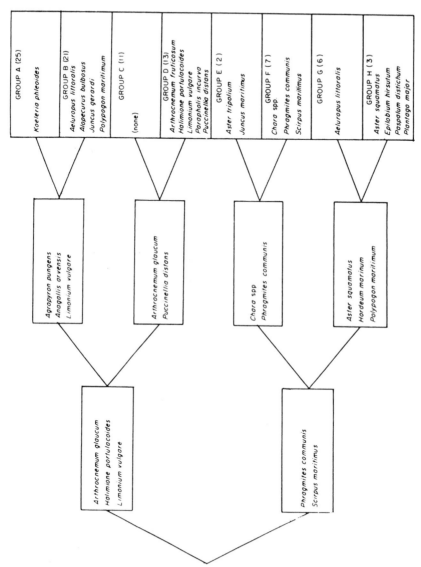

Fig. 5. The dendrogram produced by indicator-species analysis, showing both the indicator species at each successive dichotomy and the number of stands in each of the eight final groups. (From Bassett 1978, courtesy of Blackwell Scientific Publications.)

Table 3. Species Showing Significant Positive Association with the Different Vegetation Groups Recognized by Indicator-Species Analysis

	A_1	A_2	B	C	D	E/F	G	H
Brachypodium phoenicoides	***	+	·	·	·	·	·	·
Juncus maritimus	***	+	+	·	·	+	+	·
Phillyrea angustifolia	***	+	+	·	·	·	·	·
Aristolochia rotunda	***	·	·	·	·	·	·	·
Dactylis glomerata	**	**	·	·	·	·	·	·
Anagallis arvensis	*	**	+	·	·	·	·	·
Halimione portulacoides	·	***	+	·	*	·	·	·
Limonium vulgare	+	***	**	·	+	·	·	·
Bromus mollis	·	***	+	·	+	·	+	·
Koeleria phleoides	·	***	·	·	·	·	·	·
Bromus madritensis	+	***	+	·	·	·	·	+
Plantago coronopus	·	***	+	·	·	·	+	·
Agropyron pungens	+	***	+	·	·	·	·	·
Bupleurum semicompositum	·	***	+	·	+	·	·	·
Limonium bellidifolium	·	***	+	·	+	·	+	·
Crepis vesicaria	·	***	+	·	·	·	·	+
Plantago lagopus	+	***	·	·	·	·	·	·
Evax pygmaea	+	**	·	·	·	·	·	·
Catapodium rigidum	·	**	+	·	·	·	·	·
Centaurium tenuiflorum	+	*	+	·	·	·	+	·
Carex chaetophylla	+	*	+	·	·	·	+	·
Parapholis incurva	·	*	+	·	+	·	·	·
Suaeda vera	·	*	+	·	+	·	·	·
Artemisia caerulescens	·	*	·	·	·	·	·	·
Polypogon maritimum	·	+	***	·	+	·	+	*
Aeluropus littoralis	·	·	***	+	·	·	**	·
Alopecurus bulbosus	·	·	***	·	·	·	+	·
Juncus gerardi	·	·	***	·	·	·	+	+
Arthrocnemum fruticosum	·	+	*	+	+	·	·	·
Hordeum marinum	·	+	*	·	+	·	+	+
Arthrocnemum glaucum	·	+	+	***	***	·	·	·
Puccinellia distans	·	+	+	+	***	·	·	·
Scirpus maritimus	·	·	·	·	·	***	***	**
Phragmites communis	·	·	·	·	+	***	+	·
Chara spp.	·	·	·	·	·	***	·	·
Aster tripolium	·	·	·	·	·	+	***	·
Tamarix gallica	·	·	·	·	·	·	***	·
Salsola soda	·	·	·	·	·	·	***	·
Aster squamatus	·	·	+	·	·	·	·	***
Epilobium hirsutum	·	·	+	·	·	·	·	***
Paspalum distichum	·	·	·	·	·	·	·	***
Plantago major	·	·	·	·	·	·	·	***

Source: from Bassell 1978, courtesy of Blackwell Scientific Publications.
Note: ***, $p < 0.001$; **, $p < 0.01$; *, $p < 0.05$; +, $p > 0.05$; ·, species absent in that group; the other species recorded in the pasture showed no significant associations with any group.

Group A_1 was a coarse grassland dominated by *Brachypodium*, while A_2 was a species-rich grassland with many halophytes. Group B was also species rich but was notable for the hygrophilous *Juncus gerardi* and *Aeluropus littoralis*; this group contained the plots in the shallow marshes. Group C occurred on the salt flats and contained only four species, of which two were *Arthrocnemum*; Group D was rather similar but contained more species and less bare ground. The merged Group E/F occurred in the deep-water marsh and were dominated by reeds (*Scirpus* and *Phragmites*), while Group G, ecologically rather similar to B, occurred at the edge of this marsh. Finally, Group H was restricted to the Fallow fields and contained both hygrophilous plants (e.g., *Scirpus maritimus*) and many ruderal species (in the genuses *Aster* and *Epilobium*). Floristically, its closest affinities were with the marshes.

Ordination of the stands, using correspondence analysis (Hill 1974), reflected closely the groupings obtained by indicator-species analysis: Axes 1 and 3 contained most of the information used to divide the stands floristically (Bassett 1978). Axis 1 separated the stands in the marshes and Fallow fields, which had been flooded artificially until 1974; variation along the third axis corresponded with an increase in the tolerance of the vegetation to soil salinity.

The nature of the vegetation in the study area was placed in a broader, biogeographical context by achieving a synthesis with the phytosociological method, which has been used most widely in the Mediterranean. Bassett (1978) found that a fair proportion, 41 out of 110, of the species recorded were "faithful species" of four of the Braun-Blanquet orders, *Salicornietalia, Juncetalia maritimi, Thero–Brachypodietalia*, and *Phragmitetalia*. The first two are among the most important Mediterranean halophytic orders; the *Thero–Brachypodietalia* is typically Mediterranean, but a Garrigue rather than a halophytic order; and the *Phragmitetalia* is a wetland order with a wider distribution around the North African and European coastlines as far north as Scandinavia (Bassett 1978). There are two main conclusions of this work: first, the vegetation shows affinities with four vegetation types which are widespread in the Mediterranean region and are very diverse, ranging from the wetland to the dry garrigues, though most of the area is covered by halophytic vegetation. Secondly, the occurrence of the eight vegetation groups in the study area is largely controlled by two environmental factors—the water regimen and soil salinity.

Productivity

Few measurements of primary production have been made in the Camargue itself, but a great deal of work has been done elsewhere in the Mediterranean area as part of the IBP[b]. Wetland vegetation has the most productive herb layer, with maximum values for aboveground Net Primary Production (ANPP) being of the order of 3,500 gDM/m²/yr in *Phragmites*

Table 4. Above-Ground Net Primary Production for Ungrazed Vegetation in the Mediterranean Region

Vegetation type	ANPP (g/m²/yr)	Area	Source
Phragmites	600–3,700	Mediterranean	1
Scirpus lacustris	800–3,100	Mediterranean	1
Scirpus maritimus	200–1,400	Mediterranean	1
Scirpus/Phragmites	1,100	this study	
Halophyte groups	500–1,000	Camargue	2
Pastures	200–600	Mediterranean	3
Garrigue	300–400	Mediterranean	4, 5

Sources: (1) Kvet and Husak (1978); (2) Berger et al. (1978); (3) Kalinowska and Mochnacka-Lawacz (1976); (4) Lossaint and Rapp (1971); and (5) Margaris (1976)

reed beds (Kvet and Husak 1978). Using the same method (peak biomass, Appendix 4), the marshes in this study area produced an estimated 1,100 gDM (dry matter)/m²/yr.

Halophytic vegetation has received much less attention, but two measurements of production have been made in the Camargue on a type of vegetation (*Salicornietum fruticosae*) which corresponds approximately with the most extensive Facet (5) in the study area. The ANPP was 950 and 1,000 gDM/m²/yr in an exceptionally humid summer (Berger et al. 1978). These authors suggest that the values of ANPP normally range between 500 and 1,000 gDM/m²/yr.

Unimproved grasslands in this area produce between 200 and 600 gDM/m²/yr (Kalinowska and Mochnacka-Lawacz 1976). The results of the IBP production studies and of some others are summarized in Table 4.

The relatively high values of ANPP in the halophytic vegetation are surprising. It must be concluded that though the salinity of the soil has a dramatic effect on the species composition of the vegetation, it leaves productivity unaffected, at least until the salinity is so high that plant cover is reduced, as in Facet 4.

The measurement of primary production is not straightforward unless litter formation during growth is negligible, as after the dry season fires in the tropics, or in a temperate reedbed. When growth and litter formation occur simultaneously, as in the majority of temperate plant communities, then both parameters must be measured in order to estimate the ANPP. In this study, the productivity of the reedbeds was measured, using the maximum biomass method (Kvet and Husak 1978, Linthurst and Reimold 1978). Elsewhere, the plant growth which is produced on exclosure plots from which the aerial plant parts (and thus potential litter) had previously been clipped was measured at monthly intervals as an index of ANPP.

---

[b] International Biological Programme of UNESCO.

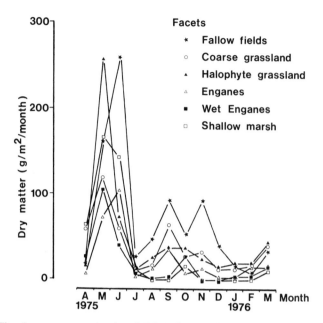

Fig. 6. The dry matter production (gDM/m²/month) on clipped plots.

The mean weights harvested from the preclipped samples in each facet are shown in Fig. 6. For the Shallow marsh, Fallow fields, and Wet enganes facets, there was a sharp peak of growth in May and June, and a slight one in the autumn, with no growth at all in the winter months. The biomass data from the Deep marsh showed that growth occurred from March to July (Fig. 7), and there was no growth in winter. In the higher-ground facets (5–7) and in the Fallow fields, however, some growth occurred in all months. Peaks were observed in spring and autumn, and there was little growth in the summer months.

To summarize, in spite of the rather unusual species composition of much of the Camargue's vegetation, the ANPP is likely to be in the upper part of the range for a Mediterranean climate. The area should therefore, in principle, be able to support relatively high densities of large herbivores. Whether this happens will depend on many other factors—both abiotic ones such as climate (cf. Andrewartha and Birch 1984) and biotic ones such as disease and predation.

A crucial factor will be the seasonality of production because strongly seasonal environments support lower densities of animals. The Camargue is highly seasonal: plant growth ceases both in the winter and in the summer drought in some vegetation groups. The horses are therefore confronted with patterns of growth which are neither simple nor similar in all facets.

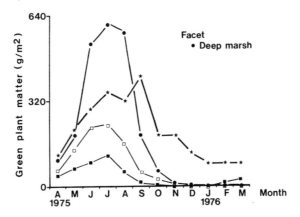

Fig. 7. Seasonal changes in the density of green herbaceous plant matter in the marshy facets. For other symbols, see Fig. 6. (From Mayes and Duncan 1986, courtesy of E. J. Brill.)

Quantity and Quality of the Herb Layer

The herbage available to the horses was measured as described in Appendix 4 ("OUT" samples) using the parameters:

1. Quantity
   - Green grass (g/m^2)
   - Brown grass (" ")
   - Green forb (" ")
   - Brown forb (" ")
   - Perennial forb (*Halimione*) (g/m^2)
   - Average height and cover (%)
2. Quality
   - Energy (Mcal/g)
   - N (%)
   - P (%)
   - Acid detergent fiber (i.e., cell walls, ADF, %).

The seasonal changes in green phytomass in the different facets are shown in Figs. 7 and 8. As might be expected from the results above, green plant matter is both much more abundant and more seasonal in the marshy facets than in the others, with the exception of the Fallow fields, which were similar to the marshes in the abundance of green matter yet were similar to the higher ground in terms of the presence of some green matter all through the year.

The seasonal changes in total plant biomass in the herb layers of the marshy facets are shown in Fig. 9. Available food for the horses in the Deep marsh changed sharply between summer and winter; in the other

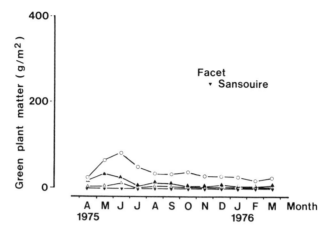

Fig. 8. Seasonal changes in the density of green herbaceous plant matter in the higher facets. For symbols, see Fig. 6 (from Mayes and Duncan 1986, courtesy of E. J. Brill).

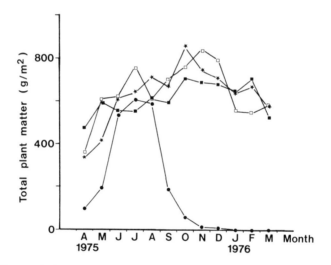

Fig. 9. Seasonal changes in density of the total herbaceous plant matter (dead reeds excepted) in the marshy facets. For symbols, see Figs. 6 and 7.

facets, however, there was a general increase through the growing season and a decrease during the winter. In the higher facets (see Fig. 10), a rather different pattern was found: there were two peaks, one after each of the periods of production shown in Fig. 6. Comparison of Figs. 7 and 8 with Figs. 9 and 10 shows that the green matter formed only a small proportion of the total phytomass of the facets on the higher ground (c. 5–15%).

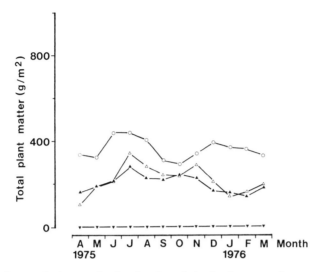

Fig. 10. Seasonal changes in the density of the herbaceous plant matter in the higher facets. For symbols, see Figs. 6 and 8.

To summarize, during this period, the plants in the marshy facets (apart from the Fallow fields) provided large amounts of green phytomass between March and October only, but those of the higher ground grew in every month of the year. Rates of growth were highest in the spring and in autumn. Green matter was most abundant in spring and summer in the marshy facets but descended to zero in the winter. In the Fallow fields and the higher ground, the seasonal changes were less marked, and some green matter was available throughout the year.

Total herb-layer biomass in the marshy facets and the Fallow fields showed a unimodal seasonal pattern, peaking at the end of the warm season; on the higher ground, it was a bimodal pattern, peaking after each of the growing periods. In all except the Deep marsh and Fallow fields, green matter made up but a small part of the total herb layer.

Between 1976 and 1983, the total phytomass in the higher facets in January declined from over 500 t (metric tons) to less than 200 t (analysis of variance [ANOVA], $F_{2,5} = 15.1$, $p < 0.01$), and in the last year, the facet with the highest phytomass was the Fallow fields, with 170 g/m².

As in most grazing lands, there was a strong negative correlation between the protein and fiber concentrations of components of the grasses[c]; see Fig. 11. The results of the analyses for *pasture quality* are therefore

---

[c]The forb *Halimione* contains considerably less fiber per unit protein than do grasses.

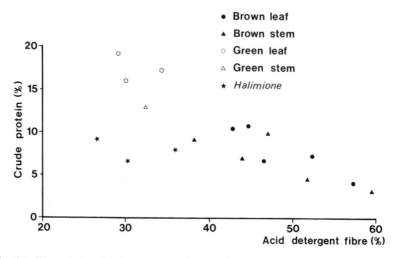

Fig. 11. The relationship between crude protein and acid detergent fiber (ADF) in different components of the herb layers, in March 1975; Facets 1, 2 + 3, 5, 6 + 7.

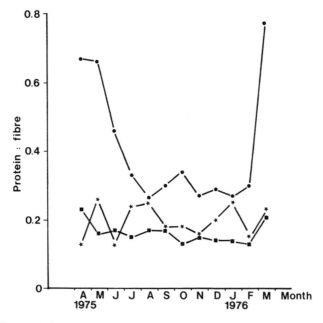

Fig. 12. Seasonal changes in the P:F ratio for the marshy facets. For symbols, see Figs. 6 and 7.

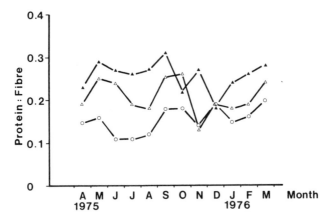

Fig. 13. Seasonal changes in the P:F (ADF) ratio for the higher facets. For symbols, see Fig. 6.

expressed as a protein/fiber (P/F) ratio. In the Deep marsh, the quality of the reed shoots (see Fig. 12) showed very marked seasonal changes. The crude protein concentrations rose in spring, and the P/F ratio was higher than that observed in the other swards. In Facets 2 and 3, the same kind of seasonal changes were observed, but of a very much smaller amplitude, while in the Fallow fields, the ratio showed no clear seasonal trends.

The results for the higher facets (5, 6, and 7) are shown in Fig. 13. Here again, there seemed to be a trend for higher values to occur in the spring (and in the autumn), but the amplitudes were slight.

For growth and lactation horses require protein levels above 12% (NRC 1978). The growing shoots of the marsh plants were always near or above this threshold (11–24%), but the whole swards of the other facets never exceeded 12% crude protein. The Camargue swards are, with the exception of the marsh plants, of such poor quality as to be limiting for growth in horses.

The lack of marked seasonal changes in the food values of the total swards is a consequence of the fact that, apart from in the Deep marsh, the green matter made up only a small proportion of the vegetation. Thus, although the whole sward may have had a low and rather constant food value, its constituent parts were extremely diverse; in Fig. 11, it can be seen that at any one time (in this case, March), protein values varied between 3 and 19%, and fiber values between 28 and 60% for different components.

In conclusion, the swards in the study area were generally of poor quality. They showed marked seasonal changes, and in any one season, there were great variations among the swards of different facets, both in quantity and in quality. These temporal and spatial variations presented the horses with choices which could affect the growth, and thus survival, of their offspring.

Animals of the Camargue

The large-mammal fauna of western Europe was never as rich as the present day African fauna, but two kinds of elephant, a mammoth, two rhinos, and a hippo species, as well as bison, wild horses, and cattle lived in western Europe in the last 100,000 years, and they probably became extinct because of overhunting by humans (Owen-Smith 1988). The last elephants, which lived on Mediterranean islands, died out less than 10,000 years ago. Even the deer (Cervidae) have disappeared from the Camargue; only the Wild boar has survived.

The bird communities of the Camargue are rich and diverse (330 species, per Blondel and Isenmann 1981) and have attracted attention to the region. With the delta of the Danube and the estuary of the Guadalquivir in Spain, the Camargue is one of the main wintering areas for Palearctic ducks and coot (Rüger et al. 1986). It hosts populations of all nine species of *Ardeidae* (herons and egrets), of which eight breed there, and a flamingo colony which has produced over two thirds of the fledged young that have been produced by Mediterranean population in the last 15 years.

The invertebrate communities are also species rich but have been studied less (but see Bigot 1961, Aguesse 1960). Bigot's work shows that the ecologically dominant species, at least in the 1950s, were earthworms (detritivores) and that the herbivorous species do not occur at high densities. Today, the land snail, *Theba pisani*, is very common, and in wet years, the impact of this species could be considerable. Unfortunately, no direct measures are available; however it is likely that it is among the mammalian herbivores that the important competitors of the horses are to be found.

Potential Competitors of the Horses

Wild boar and Wood mice feed largely on roots, seeds, and animal matter (Dardaillon 1986 and M. Jamon personal communication); they are unlikely to compete directly with the horses. Cattle and sheep were excluded from the study area but may well compete with horses where they occur together. Voles (*Microtus agrestis*) are common in coarse grasslands from which cattle and horses are excluded, around the study area (M. Jamon personal communication). Inside the study area, there was no evidence for their presence (galleries etc.), nor were any caught in 500 trap-nights. It is therefore possible that the large ungulates modify the structure of the habitat, to make it unsuitable for the voles.

The major mammalian grazers other than the horses have all been studied in detail on the Tour du Valat: coypu by Kohli (1980, 1981), rabbits by Rogers (1979, 1981; Vandewalle 1989); and grasshoppers by Boys (1981). On the basis of these studies, rough estimates of food offtake have been

Table 5. Estimated Offtake of Herbaceous Plants by the Main Herbivores in the 335-ha Study Area

Species		Total number	Mean biomass (kg/ha)	Intake (kg_{DM}/d)	Wastage (%)	Offtake (kg_{DM}/ha/d)	Total offtake (t_{DM}/day)
Coypu[1]	1977–1978	50	0.61	0.065	32	0.05	0.017
Grasshoppers[2] (dry weight)	1977	—	0.06	0.16	600	0.06	0.020
Rabbits[3]	1976	838	3.323	0.067	15	0.26	0.087
	1977	670	2.655	0.067	15	0.20	0.067
	1983	3,450	13.6	0.067	15	1.05	0.352

Sources: Data from E. Kohli[1], Jan 1979, Boys (1981)[2], and P. M. Rogers[3] (personal communication).

made (see Table 5); they will be compared with the estimates for horses in Chapter 8. Among the smaller mammals, rabbits were by far the most important primary consumers during these years. In the later, drier years of the study, their numbers increased even further, and they were clearly the main potential competitors of the horses.

Parasites—Secondary Consumers of the Horses

Wolves disappeared from the Camargue in the 1840s; since then, the only secondary consumers of the horses have been humans and the horses' parasites: blood-sucking ticks, leeches, Diptera (midges, mosquitoes, and horseflies), and internal parasites. These organisms carry many types of viral and bacterial diseases (Foil 1989) from which horses suffer. Arboviruses (Mengin 1980) such as equine encephalitis (Panthier et al. 1966), and anthrax (Poulle 1817) are known in the Camargue, but the only disease noted during this study was a single epidemic of equine influenza in 1979.

In this section, I review the current parasites, giving for each major group information on the season and time of day when they are active, and where possible, an indication of the importance of their impact on the hosts. Horseflies (Diptera, Tabanidae) are examined in more detail: though they are not as abundant as mosquitoes, their relatively long period of activity (7–20 hr) and large size mean that these insects are a scourge of both horses and cattle in the Camargue.

Many species of internal parasites, mainly Diptera and nematodes, occur in Camargue horses. The larvae of the fly *Gasterophilus* (Diptera: Oestridae) do not divert much energy from the host, nor do they cause serious lesions to the wall of the stomach (Evans et al. 1977). The roundworms (Nematoda: Strongyloidea, Trichostrongyloidea, and Ascaroidea), however, are of great economic and ecological importance. Their infective larvae occur in large numbers in pastures (e.g., 3,000 per kg of herbage, J.L. Duncan 1974) and are ingested while the horses feed.

The larvae of all the roundworms damage the internal organs of the host, especially the wall of the stomach and intestine, and the lesions they cause can lead to perforation of the gut, peritonitis, and death. The ascaroids and large strongyles (>1.5 cm, the subfamily Strongylinae) have complex migrations in the blood vessels and soft organs of the host, which they can damage severely.

The small strongyles (subfamily Cyathostominae) do not migrate beyond the wall of the gastrointestinal tract; as adults, they live on particles of food of the host, on the microflora of the gut mucosa, and they cause no damage to the wall. However, the lesions caused by the larvae and by the adult Strongylinae and *Parascaris* lead to loss of mobility in the intestine (Bueno et al. 1979), loss of appetite and lower weight gains (J.L. Duncan and Pirie 1974). Bueno et al. suggest that the loss of appetite may be a direct consequence of the reduced ability of the intestine wall to prop-

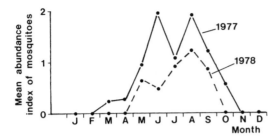

Fig. 14. Mean abundance index of mosquitoes that settled on one side of a sample horse. Index: 0 = none; 1 = 1–3; 2 = 4–10; 3 = 11–30; 4 = >30. Observations were evenly spread between sunrise and sunset.

el food. In some cases, the ulceration is so severe that it leads to perforation of the gut, peritonitis, and death.

In this study the numbers of adult nematodes in the animals were not known, but fecal egg counts showed that they carried considerable parasite burdens, both including both large and small strongyles. The eggs occurred at densities similar to those recorded in feral horses (Appendix 5.1). No deaths were caused to animals in this study, but there may have been other costs for which no direct measures are available.

The commonest tick species are *Ripicephalus* spp. and *Dermacentor marginatus*. These parasites settle at the roots of the long hairs of the horses' manes and tails, and, occasionally, in the perianal region. They were not often seen on the horses in the study herd, and careful searches during the peak of tick abundance of seven horses living in similar circumstances revealed between 0 and 18 ticks per horse ($\bar{x} = 8.2$). None were found during the 7 months when the ticks were inactive. The actual offtake of blood by ticks must therefore be trivial, but indirect costs to the horse (infection by disease organisms, development of sores at attachment sites) may be important.

There are 24 species of mosquitoes (Culicidae) in the Camargue, of which the commonest are members of the genera *Aedes*, *Anopheles* and *Culex* (Rioux and Arnold 1955). The seasonal cycle of these is extremely long, and active adults have been recorded in January (Rioux and Arnold 1955). In most years, species is a strong peak of activity in summer and autumn; see Fig. 14.

Mosquitoes have two marked peaks of activity during 24 h, at dusk and at dawn; see Fig. 15. Numbers at other times of the day are low. Therefore, any influences of these biting insects will occur at dawn and dusk. To my knowledge, no estimates have been made of the offtake of blood from large mammals by mosquitoes; this could be considerable. Mosquitoes are known to carry a wide variety of diseases (e.g., equine encephalitis, myxomatosis; Rioux and Arnold 1955), and the extremely high levels of "com-

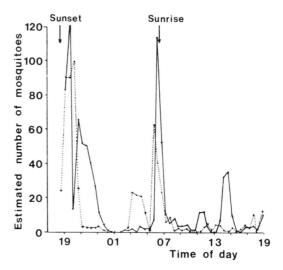

Fig. 15. Estimated number of mosquitoes settled on a sample horse. Solid line indicates September 23–24, 1975; dotted line indicates September 21–22, 1975; hours are GMT (Greenwich mean time) + 1.

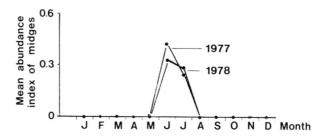

Fig. 16. Mean monthly abundance index of midges on one hindquarter of a sample horse. Index values and timing of observations as for Fig. 14.

fort behavior" shown by horses when attacked by mosquitoes leave little doubt that these insects cause distress as well as energy losses to their hosts.

The least well studied of the biting insects are the midges (Ceratopogonidae): their imagos are small (c. 1 mm long), diurnal and have a short season of activity (Fig. 16, Rioux et al. 1968), which coincides with the peak months for horseflies (see the following paragraphs).

With 24 species (Raymond 1978), the Camargue horsefly fauna is rich. Estimates of the quantity of blood taken by horseflies vary. A female can remove 20–200 μl in a meal; during the peak of fly activity, a large animal

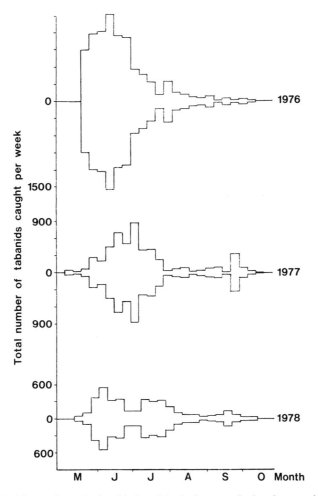

Fig. 17. Weekly catches of tabanids in a Manitoba trap during 3 years (data from Hughes et al. 1981.)

such as a cow or a horse can lose hundreds of centiliters in a day (Tashiro and Schwardt 1953). In addition, tabanids carry many types of bacterial and viral disease organisms to which horses are susceptible (Chvala et al. 1972).

Tabanids have been studied in detail in Europe (Chvala et al. 1972) and in the Camargue (Raymond 1978, Hughes et al. 1981, and Appendix 5). The larvae live in damp soil or in marshes. The blood-seeking females, which live for about 40 days, are active between May and October (Fig. 17), from sunrise to sunset (Fig. 18). The imagos are ubiquitous but show a

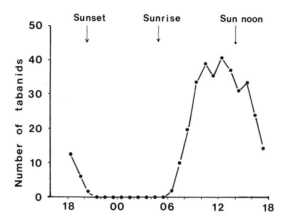

Fig. 18. Hourly arrival rate of tabanids at a Manitoba trap; means for 6 days, in June and July of 1976; hours are GMT + 1 (Data from Hughes et al. 1981.)

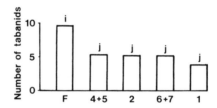

Fig. 19. The mean numbers of horseflies on a riding horse in the different facets, over all species and months (F = fallow fields; 4 + 5 = sansouïre and enganes; 2 = swallow marsh; 6 + 7 = grasslands; 1 = deep marsh). Means with the same superscript (i, j, k) are not significantly different (p = 0.05; from data in Hughes et al. 1981).

marked preference for one of the land facets, the Fallow fields (Fig. 19). They prefer colored horses to white ones (Fig. 20) but do not discriminate among individuals on the basis of sex, age, or rank (Appendix 5). The horses respond to attacks with "comfort behaviour": vigorous movements of the tail, limbs, and head, shaking their bodies, or rolling on the ground (Fig. 21 and Hughes et al. 1981).

The pattern of abundance of the flies around the two groups of horses in 1976 is shown in Fig. 22. It was broadly similar to the pattern of catches in the Manitoba trap (Fig. 18), but the bachelor herd had many more flies per individual than the breeding group. By grouping into large herds, the animals can reduce the rates of attacks (Duncan and Vigne 1979).

The pattern of change in horsefly numbers during the day was generally similar between days and herds: there was an initial buildup, followed by a period of stabilization, though the latter phase was less well marked in the

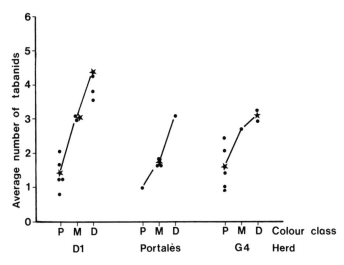

Fig. 20. The effect of color on the number of tabanids attacking a horse (circles), with class averages (stars). The results for three herds are shown: P = pale roan or grey; M = medium roan or grey; D = dark brown or black.

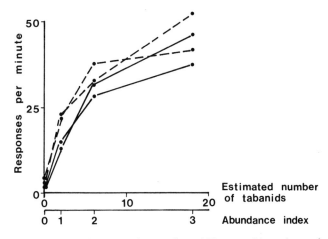

Fig. 21. The relationship between the number of flies attacking a horse (x axis) and the number of responses per minute (y axis); solid lines = females horses; dotted lines = male horses. (Data from Hughes et al. 1981.)

bachelor herd. The period of stabilization was associated with a move by the horses away from the feeding areas and onto the bare-ground resting area. Results from the previous day showed that the numbers of flies dropped to zero at about sunset. The effects of these important parasites on the horses' behavior is explored throughout this book.

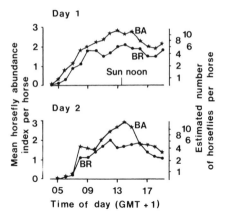

Fig. 22. The numbers of horseflies attacking horses in the two herds (BA = bachelor herd; BR = breeding herd) during the daytime in June 1976. (Redrawn from Hughes et al. 1981, courtesy of CAB International.)

Conclusions

For large mammals such as horses, the Camargue has a benign climate. The grazing is patchy, but the large wetland geophytes (especially *Phragmites*) provide food of good quality in the warm season. In the cool season, the swards are too coarse to cover the requirements of growth or lactation unless the animals are extremely selective, both feeding in the best patches and choosing the highest-quality parts (green leaves) within these patches. All the wild large herbivores are extinct, but the rabbit is a potential competitor. Though the horses have no predators other than humans, the horseflies are abundant enough to have important effects on the horses through blood loss and disease transmission.

Figure 11. A Deep marsh—lightly grazed. (P. Duncan)

Figure 12. The same marsh—heavily grazed. (P. Duncan)

Figure 9. Impact of grazing: in the Deep marsh it reduces the height. (P. Duncan)

Figure 10. On higher ground it is the perennial grasses which are reduced. Daisies proliferate on the grazed parts. (P. Duncan)

Figure 7. Fighting stallions in 1977. (J.R. Skinner)

Figure 8. (C. Feh)

Figure 5. The herd in 1976. (J.R. Skinner)

Figure 6. In 1980 after breakup into several harems. (P. Duncan)

Figure 1. Horses in marshes on Tour du Valat. (C. Feh)

Figure 2. (P. Duncan)

Figure 3. Mare and foal on coarse grassland/enganes edge. (P. Duncan)

Figure 4. Young male in the deep marsh. (C. Feh)

Figure 13. Camargue horses working cattle. (L. Violet)

Figure 14. Field observations. (P. Duncan)

Section B   Harvesting Nutrients

Chapter 3　Extracting Nutrients from Plant Tissues

The advantage of the ruminants is obviously related to the possession of pregastric fermentation in a reticulo-rumen.

Peter Van Soest
"Nutritional Ecology of the Ruminant"

When nutrients such as protein and energy are available in excess of maintenance requirements, animals use them for growth, reproduction, and building up body reserves. An ungulate which assimilates nutrients faster than its competitors will, all other things being equal, survive better and leave offspring which are more numerous or of better quality than those of its competitors. It will, therefore, be more successful in evolutionary terms.

In this chapter, I review what is known about the factors affecting food intake and digestibility in equids, compare them with ruminants, and test the predictions of the nutritional model shown in Chapter 1, Fig. 5.

The rate at which an individual extracts nutrients[a] is a function of the daily intake, the true digestibility, and the metabolic losses of the nutrients. In most digestion trials, apparent, not true, digestibility is measured (apparent digestibility is the true digestibility less metabolic fecal losses—mucus and epithelial cells from the gut wall). Because equids have a higher rate of fecal loss than ruminants (Axelsson 1941), the apparent rate of intake is therefore a more meaningful measure for comparing extraction of nutrients and is used for comparisons between equids and other herbivores in this book. Digestibility may differ among nutrients, but the values are

---

[a] Synonymous with the daily intake of apparently digestible dry matter.

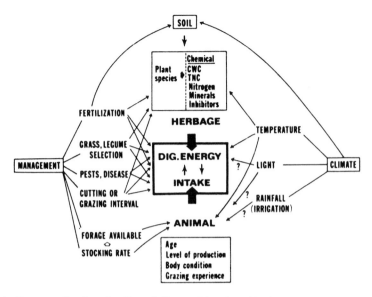

Fig. 1. Factors affecting the digestibility and intake of herbage by grazing animals. (From Reid and Jung 1982, courtesy of CAB International.)

closely correlated with the digestibility of the whole diet (dry matter), so this review focuses on voluntary intake of dry matter per unit of metabolic weight (VI, kg DM/W 0.75/day) and on the apparent digestibility of dry matter (DDM, %).

The extensive work which has been carried out on food intake and digestibility in ruminants, principally cattle and sheep, has shown that there are two main classes of determinants: (1) animal factors, and (2) forage (or herbage) factors; see Fig. 1 (Church 1979, Van Soest 1982 and Hacker 1982).

Digestibility (DDM) varies across animal species. Cattle digest high-fiber roughages to a slightly greater extent than smaller animals such as sheep, and to a very much greater extent than browsers (Hofmann 1973, 1989; Van Soest 1982, Table 20.7; Prins et al. 1984).

Among ruminants, the considerable differences among species in their ability to digest roughages are caused principally by differences in gut morphology and in body size. Within species, however, animal factors such as level of production have little effect on DDM.[b] Forage factors (especially fibrousness), on the other hand, affect DDM strongly: cattle can digest 80% of high-quality forages, but only 50% of very low-quality foods.

---

[b] Digestibility values obtained with wethers (castrated male sheep) are used for calculating the rations of lactating ewes.

Intake rates are more difficult to predict because within a species, they are affected by both animal and forage factors. The ultimate determinant of food intake is the animals' nutrient requirements: animals try to eat to balance their requirements (Baile 1979). For a fixed level of nutrient requirements, ruminants eat more as the digestibility of food declines. However, the capacity of the alimentary canal is quickly filled. For instance, Conrad et al. (1964) showed that in dairy cows, intake is determined by requirements only when the diet digestibility was above a certain threshold (in their study, 67%). Below the threshold, intake declined with declining food quality and the animals could not balance their nutrient requirements with their nutrient intake.

On these coarser forages, the animals' intake is determined by the rate at which digested food leaves the alimentary canal. Numerous studies of ruminants have shown that the retention time of a forage depends principally on its fibrousness (Campling and Lean, 1983). Plant factors are therefore key determinants of the digestibility and intake of the coarser forages typically eaten by free-ranging ruminants.

Comparing the Quality of Forages

Two chemical methods of characterizing the quality of forages are commonly used: (1) the traditional system Weende (or proximate) analysis and (2) the Detergent System.

The Weende system divides forages into

- Crude fiber
- Crude protein
- Ether extract
- Nitrogen-free extract (NFE)
- Ash

These constituents are not chemically consistent, and the newer detergent system has been developed in the past 25 years (see Van Soest, 1982, for a full account). The main constituents of the detergent system are outlined in Table 1.

In detergent analysis, boiling the forage in neutral detergent leaves a residue (neutral detergent fiber, NDF), which is the plant cell wall, less the pectins. Pectins are dissolved with the cell contents. Correction for the loss of pectin is not normally made, and NDF is commonly equated with the cell-wall fraction, which ranges between about 35 and 80% in forages.

Hydrolysis of the NDF in acid detergent leaves a residue (acid detergent fiber, ADF) which is cellulose, lignin, cutin, silica, and indigestible nitrogen; see Table 1. Hemicelluloses and fiber-bound protein are dissolved. In fibrous feeds, both the amount and the lignification (Lignin index,

Table 1. Digestibilities of Different Components of Forage: Based on the Detergent System

Origin		Class	Chemical composition	Digestibility
Cell contents	Neutral detergent solubles	1	Soluble carbohydrates Starch Lipids Organic acids Protein Pectin Other water soluble matter	>90% or near total availability
Cell walls (neutral detergent fiber)	Acid detergent solubles	2	Hemicellulose, fiber-bound protein	Variable; depends on ligninization, silicization, and cutinization
	Acid detergent fiber	3	Cellulose Lignin Cutin Silica Tannins, essential oils and polyphenols	Variable Indigestible Indigestible Indigestible May be absorbed but are not used.

Source: Data from Van Soest 1982, courtesy of O & B Books.

Lignin/ADF $\times$ 100) of ADF is high. ADF is the detergent fraction that is most similar to crude fiber.

These detergent fractions are strongly related to both forage intake and digestibility in ruminants across a wide range (40–80% NDF) of forages; see Table 2.

Factors Affecting the Digestibility of Forages in Equids

Digestibility is measured as DDM in some studies, and as the digestibility of organic matter (DOM) in others. The two measures are closely related in horses (DDM = 1.04 DOM − 0.88, Appendix 6, second section), so DOM values have been converted into DDM where necessary.

Animal Factors

It has been claimed that wild equids are better digesters of forages than domestics, but they have rarely been compared on exactly the same forage.

Table 2. Correlations between Chemical Fractions of Forages, and Their Intake and Digestibility for Ruminants

Detergent fraction	Intake	Digestibility
Neutral detergent fiber (NDF)	-0.76^{***}	-0.45^{***}
Acid detergent fiber (ADF)	-0.61^{***}	-0.75^{***}
Lignin	-0.08	-0.61^{***}
Crude protein	$+0.56^{***}$	$+0.44^{***}$

Note: n = 187; *** $p < 0.001$.
Source: Data from Van Soest 1982, Table 6.9, courtesy of O & B Books.

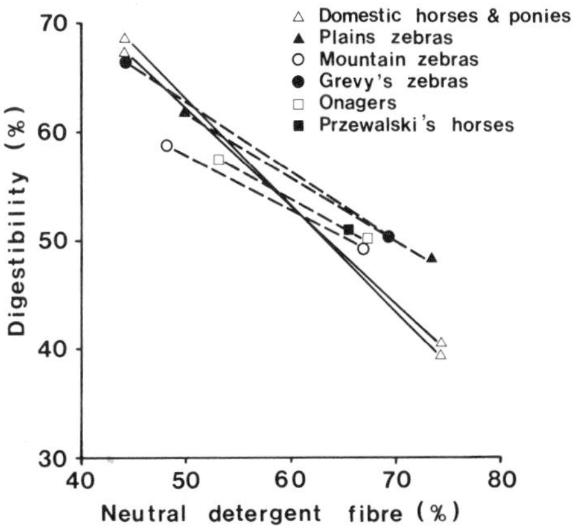

Fig. 2. The digestibility of organic matter by domestic and wild equids. (Data from Foose 1982.)

The closest comparison comes from the work of Foose (1982), who studied the digestion of alfalfa and grass hays both by five wild species and by domestic horses and ponies. The results are shown in Fig. 2. The domestic animals digested alfalfa hays better, and the wild ones did better on grasses. The regressions are not significantly different: the grass hay fed to the domestics had a particularly high lignin content, so the differences are more likely to be due to variations in the hays than to differences in the digestive physiology of the equids. This conclusion is supported by the work of Gakahu (1982), who found that Plains zebras and donkeys digested three hays made from alfalfa and a wild grass to similar extents.

Slade and Hintz (1969) showed that ponies digested forages to a greater extent than horses. Few other data have become available since then (Ralston 1984, Martin-Rosset et al. 1984). The difference, 1–2 percentage points, though perhaps important in rationing is so slight that it will not be considered further here. Gut morphology is broadly similar across species, and although the processes of mastication and digestion differ among ani-

Table 3. Analysis of the Effects of Crude Fiber and Crude Protein on the Digestibility of Forage Dry Matter in Equids

Variable	Simple r	Independent		Cumulative			Variance explained (%)
		t	p	F	df	p	
Crude protein	0.66	5.90	<0.001	41.7	1	<0.001	44
Crude fiber	−0.46	3.18	<0.01	29.4	2	<0.001	53
Trial				44.0	8	<0.001	79
Residual					53		100

Note: Data from sources quoted in Appendix 6; grass and legume forages.

mals of different sizes (Meyer et al. 1975), species and size do not have notable effects on diet digestibility in equids.

The effect of animal factors within a species, such as the level of production or stage of gestation, on the digestibility of forages has received little attention in equids (Boulot 1987).

Forage Factors

The most comprehensive study is that of Olsson et al. (1949), who showed that the crude fiber content was the most important determinant of digestibility over a wide range of foods (over 1,000 foods, crude fiber 2–48%). The protein content of the diet accounted for an additional part of the variations in digestibility.

This wide range of foods, which included straws, hays, grain, and by-products such as beet pulp, is eaten by domestic horses. Free-ranging equids have a diet composed almost entirely of grasses and forbs (see Chapter IV). When the analysis is restricted to forages alone, the best predictor of digestiblity is crude protein (see Table 3), though crude fiber had an additional significant effect. The two variables together accounted for about half of the variance in digestibility. In addition, there was a highly significant effect of *trial*. This means that the proximate variables, crude fiber and protein, had different effects on digestibility in different experiments, so the overall relationships are not reliable. A similar analysis was performed on a data set containing most of the data used here, together with results of experiments by Wolff in the 19th century and at the Institut National de la Recherche Agronomique (I.N.R.A.) (Martin-Rosset et al. 1984): in this data set, crude protein had no significant effect on digestibility in addition to the effect of crude fiber (see Table 4, Equation 4).

These analyses show that forage factors have important effects on digestiblity in equids, but these are inconsistent across experiments. The likely explanation is that these variations are due to inconsistencies in the chemical composition of the proximate variable "crude fiber."

Table 4. Relationships between Digestibility and Forage Composition Measured by Proximate Analysis

Equation	r, R	F	df	R	Forages	Ranges
1. DDM = 42.9 + 0.962 CP	0.618	48.7	1, 79	(0.001)	Grasses, legumes; CP (20.0%)	DDM = 40–69% CP = 4.4–19.7%
2. DDM = 43.0 + 0.977 CP	0.660	41.7	1, 54	(0.001)	All except alfalfa	DDM = 40–66% CP = 4.4–14%
3. DDM = 60.9 + 0.856 CP − 0.494 CF	0.725	29.4	2, 53	(0.001)	All except alfalfa	CF = 25–43%
4. DOM = 78.3 − 0.075 CF	0.414		73	(0.000)	Grasses, legumes	CF = 23–40%

Note: DDM = digestibility of dry matter; DOM = digestibility of organic matter; CP = $N \times 6.25$; CF = crude fiber; Equations 1, 2, 3 calculated from data in the references cited in Appendix 6; Equation 4 from Martin-Rosset et al. (1984), Tableau 4.

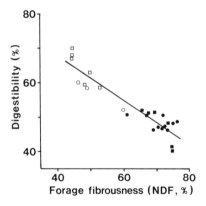

Fig. 3. The relationship between the digestibility of dry matter (DDM, %) and forage fibrousness. The regression line is given in the text, Equation 1; open points (circles and squares) = alfalfa; solid points = grass (sources of data: squares = Foose 1982; circles = Fonnesbeck 1969).

Table 5. Correlation Matrix of the Digestibility of Dry Matter of Grass and Legume Forages in Equids, and the Detergent Variables

Variable	DDM	NDF	ADF	Lignin ratio
DDM	1			
NDF	0.932***	1		
ADF	0.636***	0.589**	1	
Lignin ratio	0.649***	0.723***	0.002	1
Crude protein	0.842***	−0.835***	−0.600**	0.470*

Note: n = 25; data from Fonnesbeck 1968, Foose 1982; $*p < 0.05$; $**p < 0.01$; $***p < 0.001$

The detergent system has been used in only two studies of equids. Digestibility was closely and negatively correlated with the cell-wall fractions (see Fig. 3 and Table 5).

$$DDM = 93.3 - 0.634 \, NDF\% \qquad (1)$$
$$(r^2 = 0.872, \, N = 25, \, p < 0.001)$$
$$(\text{range of NDF} = 44\text{--}76\%)$$

None of the other constituents—ADF, lignin-ratio, or protein—had any significant additional effect on digestibility (see also Van Soest 1982, Fig. 12.8). By contrast, the digestibility of forages in ruminants is best predicted by the *summative equation*, in which NDF and the lignin ratio have additive effects (see Van Soest, 1982, Table 6.12).

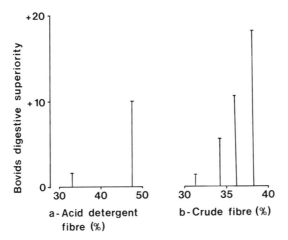

Fig. 4. Cattle digest a greater proportion of the organic matter of forages than equids, and their superiority increases with the fiber content of the forage; superiority is expressed as the difference between the bovid and equid digestion coefficients (bovid − equid). (Data from [a] Foose 1982; [b] Vander Noot and Gilbreath 1970.)

Comparative Digestion of Forages by Equids and Bovids

All the published studies show that bovids are more effective digesters of dry matter than are equids (Axelsson 1941, Haenlein et al. 1966b, Hintz 1969, Vander Noot and Gilbreath 1970, Prins et al. 1984, Martin-Rosset et al. 1984).

On good-quality forages, equids are nearly as efficient as bovids, but as the level of crude fiber or of cell walls increases, the ruminants become relatively more efficient; see Fig. 4. Forages with a higher fiber content are digested more slowly; in bovids, they are also retained for longer than good-quality forages. In equids, where passage time decreases only slightly or not at all as forage fibrousness increases, the extent to which the coarser forages are digested declines faster than in bovids.

The effect of NDF on digestibility in cattle has been calculated from data of the NRC (1984):

$$DDM = 86.6 - 0.485 \, NDF\% \qquad (2)$$
$$(r^2 = 0.355, \, N = 54, \, p < 0.001)$$

Comparative digestion coefficients for different levels of NDF are given in Fig. 5.

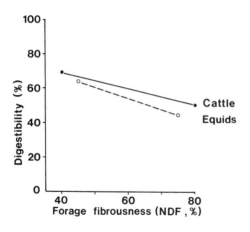

Fig. 5. Comparative digestibility of dry matter (DDM, %) in equids (Equation 1) and cattle (Equation 2), as a function of forage fibrousness.

Factors Affecting Food Intake

Animal Factors

Domestic and wild species eat similar quantities; see Table 6. Animals at different levels of production, however, eat very different amounts; see Figure 6. The geldings used for the feeding trials in the agricultural literature ate 75–115 g/$W^{0.75}$/day (median = 95 g; Haenlein et al. 1966b, Fonnesbeck 1969, Darlington and Hershberger 1968, Foose 1982, Chenost and Martin-Rosset 1985). Pregnant mares eat 85–97g/$W^{0.75}$/day, while lactating mares eat twice as much, 155–188g/$W^{0.75}$/day (Boulot 1987).

Wild-caught zebras, shortly after capture and taming, had median intakes of 167 g/$W^{0.75}$ (range 158–169 g/day, Ngethe 1976) and 171 g/$W^{0.75}$/day (range 162–191 g/day, Gakahu 1982). The results of these studies are particularly relevant to natural conditions because these animals had been raised on the range and were, with one exception, growing or breeding.

Table 6. Comparative Intake Rates of Wild and Domestic Equids

Hay	Intake (medians, g/$W^{0.75}$/d)			Difference (%) $\left(\dfrac{W-D}{D} \times 100\right)$
	Reference	Wild	Domestic	
Alfalfa	a	105	96	+9
Mature grass	a	111	104	+7
Mature grass	b	161	165	−2
Grass/alfalfa	b	174	164	+6
Alfalfa/grass	b	195	193	+1

Source: Reference a—Foose (1982), 2 domestic breeds, 5 wild species; reference b—Gakahu (1982), 1 domestic, 1 wild species.

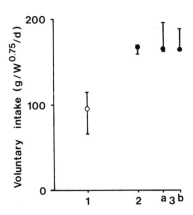

Fig. 6. Voluntary intake of dry matter (VI: median values and ranges) in different sets of equids. Non-growing, nonbreeding animals (open circles): (1) Domestic equids, 25 trials (references as for Appendix 6). Growing or breeding animals (filled circles): (2) Wild-caught zebra, 3 trials (data from Ngethe 1976); (3a) Wild caught zebra, 3 forages (3b) Asses. (Data from Gakahu 1982.)

Breeding or growing equids are therefore capable of intake rates very much higher (median = 170 g/$W^{0.75}$/day) than are shown by geldings in standard feeding trials.

Forage Factors

Given the strong effect of animal factors on intake rates, any evaluation of the effects of forage factors must be restricted to one animal type (e.g., geldings on maintenance or lactating mares). The most important type for the population is the breeding mares. Unfortunately, very few published data are available for these animals (but see Martin-Rosset and Doreau 1984b), and it is not possible to evaluate the effects of forage factors on their intake rates (Boulot 1987).

In geldings, voluntary intake (VI) ranges between 70 and 115 g/$W^{0.75}$/ day (median = 95). When feeding on concentrates, dilution of the ration with an indigestible ballast (wood shavings) caused an increase in intake

Table 7. Correlation Coefficients (ns) of Voluntary Intake (VI) by Horses, with the Detergent Variables

Variable	NDF	ADF	Lignin ratio	Crude protein
VI	−0.460	−0.275	0.393	0.131

Note: n = 16; data from Fonnesbeck 1969, Foose 1982.

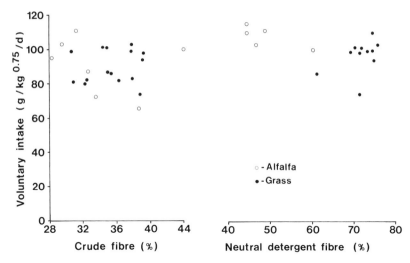

Fig. 7. The relationship between the voluntary intake of forages in equids and two measures of their fibrousness. There was no significant relationship with either measure, for alfalfa or for grasses. (Data from Darlington and Hershberger 1968, Fonnesbeck 1968, Fonnesbeck et al. 1967, Foose 1982, Haenlein et al. 1966a.)

(Laut et al. 1985). However, when feeding on hays alone (see Table 7 and Fig. 7), intake tended to *decline* with increasing fiber content, though the effect was not significant. On a larger data set, using crude fiber, Boulot (1987) found a significant, negative effect.

There is also no significant correlation between VI and the digestibility of dry matter (see Fig. 8, $r = 0.04$, $N = 26$). The pattern may in fact be less simple: among alfalfa hays (see Fig. 8), intake tended to increase as digestibility increased ($r = 0.51$, $N = 9$, n.s.). In grass hays, however, intake did not vary with increasing digestibility (Boulot 1987).

Further data are needed to check these conclusions because digestibility and trial are confounded. It is, however, known that rates of intake are affected by changes in the size of the food particles: equids consume more of a forage when it is pelleted (Fig. 9). Theoretically, this increase must be caused by an increase in one or more of the following:

1. The dry weight of digesta
2. The rate of digestion
3. The passage rate of food

The first possibility has not yet been tested. Regarding the second possibility, the increased intake of high-quality alfalfa hay could well be caused by a higher rate of digestion—this has been demonstrated in the rumen (Van Soest 1982, Fig. 13.6), and the same principles are likely to hold for equids.

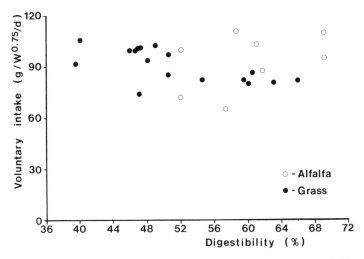

Fig. 8. The relationship between voluntary intake (VI) and the digestibility of dry matter in domestic horses. (Data from references in Fig. 7.)

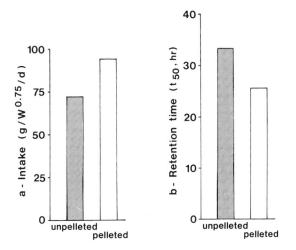

Fig. 9. The effect of pelleting on the intake and on the retention time of food in equids. (Data from Haenlein et al. 1966a; Hintz and Loy 1966.)

There is good evidence that passage rates (the third possibility) increase when foods are pelleted (Fig. 9b; and the results of Haenlein et al. 1966a imply a similar effect). Pelleting decreases particle size (Van Soest 1982, Fig. 13.5), and equids, in common with many other species of herbivores, probably retain large particles for longer than small ones (Clemens and Stevens, 1980).

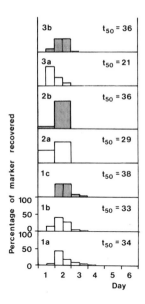

Fig. 10. Passage rates of good quality alfalfa hay (open bars) and poor quality grass hays (shaded bars) in equids; 1a = alfalfa, 1b = alfalfa and barley, 1c = timothy and barley (Vander Noot et al. 1967, horses); 2a = alfalfa, 2b = timothy (Foose 1982, wild equids); 3a = alfalfa, 3b = timothy (Foose 1982, domestic equids); t_{50} = retention time (hours).

These results provide a possible explanation for the higher intake rates on good-quality alfalfa hay: if this fragments faster or to a greater extent than other forages, then the consequent increase in passage rate would allow intake to be increased. This is exactly what was found in three comparisons between alfalfa and grass hays; see Fig. 10 (Graphs 1, 2, and 3).

The mechanism by which the larger particles are slowed down in the gut of equids is not known. Some large items are retained by the gut of equids for very long periods of time: The largest markers of Clemens and Stevens (1980) were 2-cm pieces of polyethylene tube, and more than half were retained for more than 10 days. Janzen (1981, 1982) has observed that some large seeds remained inside horses for over a month.

High rates of passage and digestion may, therefore, allow equids to increase their intakes when feeding on very high-quality leguminous forages which fragment easily. In medium- to low-quality forages, however, the available information suggests that intake is constant.

Fig. 11. Voluntary intake of (a) good-quality alfalfa hays, and (b) poor-quality grass hays, by wild and domestic species of herbivores. Open circles—equids (Przewalski's horse; Grevy's, plains, and mountain zebras; onager; domestic horse, pony); filled circles—bovids (grazers only: American and European bison; Asian, African forest, and savanna buffalo; waterbuck; gemsbok; domestic cattle, sheep); crosses—other perissodactyls (Indian, white, and black rhinos; American and Malayan tapirs); filled triangles—elephants (Asian and African). (Data from Foose 1982.)

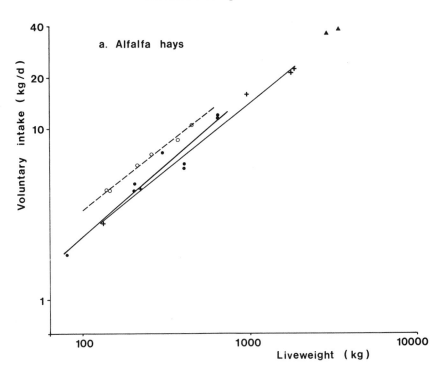

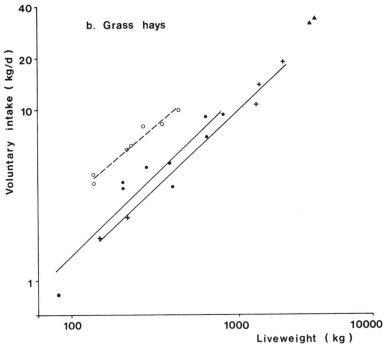

Table 8. Intake of Organic Matter by Equids and Some Other Herbivores

| | Daily Intake | | | | |
| | $(g/W^{0.75})$ | | $(\%W)$ | | |
	Alfalfa	Grass	Alfalfa	Grass	n
Grazing bovids	84[b]	62[d]	1.99	1.26	8, 9
Equids	109[a]	104[c]	2.82	2.72	6, 7
Other perissodactyls	78[b]	62[d]	1.63	1.08	5, 5

Note: median values for W = liveweight, kg; values in the same column with different super-scripts are different; $p < 0.05$; t-test; data from Foose 1982.

Comparative Rates of Intake in Equids and Bovids

A remarkable study of the use of forages by some 30 species of herbivores has been carried out by T.J. Foose (1982), using animals in zoos in the United States. The results for intake by equids and grazing bovids are shown in Fig. 11 (Graphs a and b). On alfalfa hays, equids ate slightly but significantly more per unit of metabolic weight than did either the other perissodactyls or the grazing bovids; see Table 8. On the very coarse grass hays (cell walls, NDF = 65–75%, crude protein = 5–7%) the difference be-came striking as equids maintained their levels of intake, while the bovids predictably dropped theirs. Curiously, the other monogastric herbivores (elephants and other perissodactyls, including both grazing rhinos and browsers, such as (black rhino and tapirs) showed a similar pattern to that of the bovids.

These results are consistent with those of Haenlein et al. (1966b), who showed that the daily intake of alfalfa by horses was slightly higher than cattle and much higher than sheep.

The high rates of intake in equids must come from increases in one or more of the following:

1. Weights of digesta
2. Digestion rates
3. Passage rates.

Regarding the first possibility, the amount of digesta carried by ungulates is closely determined by their body size (see Fig. 12) and increases isometrically with liveweight. The wet weight of gastrointestinal (GI) con-tents represents 13–20% of their liveweight for the range 100–600 kg, and there appear to be no differences between fore- and hindgut fermenters. This conclusion should be treated with caution because dry weights of the GI contents are more relevant to this issue than wet weights. Nonetheless, it is likely that the amount of digesta that an herbivore can carry is con-strained by its need to be agile in order to avoid predators. This is sup-

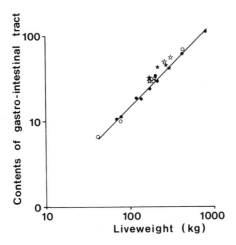

Fig. 12. The relationship between the capacity of the gastrointestinal (GI) tract and liveweight in grazing bovids (filled circles and regression), and hindgut fermenters (open circles, from Demment and Van Soest 1985, courtesy of the American Naturalist). Superimposed are data from this study for cattle (filled stars) and horses (open stars). For methods, see Appendix 6 (fourth section).

ported by the fact that digesta represent a higher proportion of liveweight in species which are less vulnerable: the two highest values come from animals with special ways of avoiding predators—hippo and domestic cattle. The available evidence, therefore, suggests that the capacity of the GI tract of ungulates is closely constrained and that equids do not carry any more digesta than bovids.

Regarding the second possibility, comparative digestion rates have received little attention, but the data available suggest that rates of digestion are not as high in horses as in cattle (Koller et al. 1978).

Regarding the third possibility, the passage rates of forages in grazing ruminants, usually measured as mean retention times (t_{50}) are difficult to quantify because liquids and solids move at different speeds, and so do particles of different sizes (Clemens and Stevens 1980). The retention time of forages has been investigated in detail because it is a key parameter in the determination of intake rates (Campling 1970; Thornton and Minson 1972; van Soest 1982, pp. 180 et seq.). In cattle particle retention times are about twice as long as in equids (e.g., 79 hr vs. 29 and 34 hr in horses and ponies, respectively; Van Soest 1982, Table 13.1). It is therefore the high passage rate of food in equids which allows them to achieve relatively high intake rates.

The median value for intake by productive equids (see page 63) is 170 g/$W^{0.75}$/day, and their intake rates do not appear to be affected by forage factors. Bovids, however, are strongly affected by both animal and

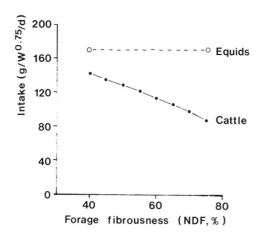

Fig. 13. Comparative rates of daily dry matter intake (DMI) in equids (page 63) and cattle (Equation 3), as a function of forage fibrousness.

forage factors: because the most productive animals consume the most, I use here the results for lactating cows. The best predictor of intake is NDF (Rohweder et al. 1983):

$$DMI = 84.7 - 3.69NDF\% + 32.37\sqrt{NDF\%} \qquad (3)$$
$$(r^2 = 0.62, p < 0.001, N = 271)$$

Comparative intake rates in the two species are given in Fig. 13.

A Model of Nutrient Extraction in Equids and Bovids

I have shown in the sections above that intake and digestibility respond differently to variations in forage fibrousness in equids and bovids. This is particularly clear for intake, which in equids appears to be relatively constant but in bovids declines sharply with increasing fiber. The relative effectiveness of the two groups at extracting nutrients will therefore vary with forage fibrousness.

I have used the relationships described above between forage fibrousness (NDF) and both intake and digestibility, to make a simple model of the relationships between the rate of extraction and forage fibrousness in the two groups. Nutrient extraction rates are measured as the daily digestible dry matter intake (DDMI). For equids the apparent rate of extraction of dry matter at a given level of NDF was calculated as the product of the digestion coefficient (calculated from equation 1), and the intake rate (Fig. 13: 170 g/W$^{0.75}$). For bovids, the data from lactating cattle (Equations 2 and 3), were used; see Fig. 14. The average rate of extraction in bovids falls from about 90 g/W$^{0.75}$ on high-quality forages (NDF = 45%) to about 40 g/W$^{0.75}$ on coarse roughages. Productive equids, eating to capacity, extract

Fig. 14. Comparative rates of nutrient extraction, measured as the daily digestive dry matter intake (DDMI), as a function of the forage fibrousness, calculated from data in Figs. 5 and 13. (From Duncan et al. 1990, courtesy of Springer-Verlag.)

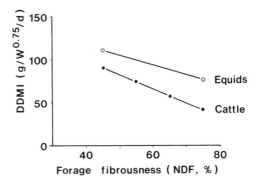

more (110–75 g/$W^{0.75}$) than bovids at all levels of forage fibrousness. The difference is particularly marked on the coarsest forages: at 75% NDF, equids extract nutrients at 175% the rate calculated for bovids.

This analysis was repeated for less productive animals, using the results of trials in which geldings and wethers were used for intake rates (95 g/$W^{0.75}$) for equids and the equation of Van Soest (1965) for sheep. The model again showed that equids assimilate nutrients at a higher rate than ruminants for forages with NDF >45%.

This model leads to two conclusions:

1. The very high intake rates of equids more than compensate for their slightly lower coefficients of digestiblity and allow equids to extract more nutrients than comparable bovids right across the range of quality encountered in natural forages.
2. The difference in extraction rates is greater on coarse than on high-quality forages.

An objection to these conclusions is that they are based on average values and that direct comparisons were not made using the same forages. So far, this has been done in three studies, only one of which has used the detergent system of forage analysis (Foose 1982).

The results for domestic animals are shown in Fig. 15 (Graph a). In this study, exactly the same diets (a good alfalfa hay and a poor grass hay) were fed to all species, and both conclusions are upheld. Equids extracted more from both diets; and the difference was smaller on the good hay than on the poor one (median values for equids are 120 and 190% of the values for bovids, respectively).

The results for wild animals are shown in Fig. 15 (Graph b). The conclusions are again upheld, with equids superior across the whole range of 45–75% NDF. Again, equids did relatively better on the most fibrous hays (120 and 145% of the bovid value at 45% and 75% NDF).

Similar results have been obtained in two other studies, though NDF values are not available here; see Table 9. In the first study, quarter-horse-

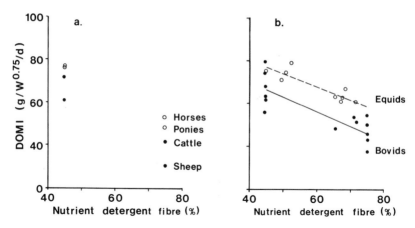

Fig. 15. Comparative rates of nutrient extraction, measured as the daily digestible organic matter intake (DOMI), as a function of forage fibrousness: (a) domestic animals, and (b) wild species held in zoos; see Fig. 11 for details. (Data from Foose 1982.)

Table 9. Differences between the Apparent Rates of Nutrient Extraction by Equids and Cattle

Hay quality	Species	Live-weight (kg)	Nutrients $(g/W^{0.75})$	Difference, % $\left(\dfrac{equid - bovid}{bovid}\right)$	Ref
Good	Horse (domestic)	487	61	+52	a
	Cow (taurine, domestic)	450	40		
Medium	Horse		71	+58	
	Cow		45		
Low	Horse		49	+48	
	Cow		33		
Good	Zebra (plains)	198	117	+46	b
	Donkey (domestic)	199	113	+41	
	Steer (Zebu, domestic)	154	80		
Medium	Zebra		95	+36	
	Donkey		91	+30	
	Steer		70		
Low	Zebra		71	+28	
	Donkey		67	+16	
	Steer		58		

Note: (a) Johnson et al. 1982, and (b) Gakahu 1982; from Duncan et al. 1990, courtesy of Springer-Verlag.

type mares and beef cows were fed three hays. In each case, the mares ate and extracted more, by 48–58%. The second compared growing male zebra and Masai donkeys with young zebu steers. Again, on all forages, the equids ate and extracted more, by 16–46%. In these studies, the equids did best relative to bovids on the best-quality hays: without data on NDF concentrations, it is not possible to compare these results with the simple model above.

General Discussion

This model, which is based on an analysis of the effects of only one forage variable—NDF—on rates of nutrient extraction, is supported by the available empirical data. It shows that equids assimilate nutrients at higher rates than comparable bovids over the range of forage quality normally used by grazing animals. It is clear that the rumen is more efficient than the GI tract of equids at extracting the most out of forages eaten, but the equid GI tract maintains a higher daily rate of nutrient extraction.

In the wild, the survival and productivity of grazers might be determined by the rate of extraction not of dry matter, but of particular limiting nutrients. An obvious candidate, especially in tropical ranges, is protein. I have not done the calculations on a protein basis instead of a dry matter basis because it is well-known that equids are able to digest protein almost as well as bovids (see Vander Noot and Gilbreath 1970, and studies reviewed there), so their superiority would be clearer still for protein extraction.

Comparable information for energy and minerals is not readily available, but there is no reason to believe that equids are sufficiently poor digesters of any important nutrient for the advantage conferred on them by their very high rates of intake to be outweighed.

This review therefore supports the explanatory model proposed by Bell (1970) Janis (1976) and Foose (1982) to explain the survival of equids in the face of competition from the evolutionarily younger grazing bovids (see page 13). However, it shows that equids are relatively more effective at extracting nutrients from all types of natural forages. Equids are far from the Miocene relics that they have been painted, surviving on the ecological refuge provided by coarse, straw-like plant tissues which are unsuitable for ruminants. Compared with the well-known grazing ruminants, equids represent an alternative evolutionary response by the ungulate order to the huge food resources available in herbaceous plants.

The different digestive system of hind-gut fermenters, with its different physiological constraints, may impose different patterns of behavior— feeding, anti-predator, and social, on the equids. These constraints may explain why it is that grazing bovids are more abundant than equids in virtually all natural ecosystems (see Chapter 1, Grazing Bovids and Equids: Competition and coexistence?).

In particular

1. The abundance of their food in natural grasslands may be inadequate for free-ranging equids to harvest for themselves the large quantities that they can consume in the feeding trials reported in this chapter.
2. The ruminant digestive system may allow bovids to exploit plant tissues that are not available to equids (e.g., because they are protected by plant secondary metabolites).
3. The need to eat such large quantities of food may require longer feeding times, and thus expose the equids to higher rates of predation than the bovids.

In the following chapters the diets and feeding behavior of the free-ranging Camargue horses will be described, compared with what is known of other equids and grazing bovids, and interpreted with these hypotheses in mind.

Chapter 4 Diets—Their Botanical Composition and Nutritional Value

Theirs is no earthly breed
Who only haunt the verges of the earth
And only on the sea's salt herbage feed
Roy Campbell
"Horses on the Camargue"

The use herbivores make of available plant matter is rarely random—large herbivores are generally selective in their feeding, preferring some and avoiding other components of the available plant matter. The choice of food plants determines the nutritional quality of the diet, and thus the rate of energy flow into an herbivore population. It also directs the animals' impact on the vegetation.

Herbivores discriminate among different species of plants (e.g., Arnold and Hill 1972; sheep, Arnold 1963; cattle, Bredon et al. 1967; wild and domestic ruminants and zebras, Casebeer and Koss 1970) and among morphological parts of plants, "parts selection," often preferring leaves to leaf sheaths and stems (e.g., sheep, Arnold and Dudzinski 1978, page 100; buffalo, Sinclair and Gwynne 1972; hartebeest, Stanley Price 1978). Within plant parts, herbivores may show strong preferences for tissue at different growth stages: in particular, it has been found that *green* leaf is a highly preferred plant component (for cattle, Reppert 1960; sheep, Arnold 1963; topi antelope, Duncan 1975).

Though selective feeding has been described extensively, there is only the beginnings of a body of theory to account for the animal's behavior in terms of its function. Early contributions were those of Ivins (1952) and Arnold (1963), while more recently, Owen-Smith and Novellie (1982) and Belovsky (1984, 1986) have reviewed the literature and provided models of selective behavior.

The fundamental parameter which determines the quantity of a plant component eaten is the quantity available, both absolutely and relative to the rest of the herbage. Below a certain level of availability, it is unprofitable for an herbivore to search for a component. At equal levels of availability, it seems that most of the variance in the use of plants by grazing animals depends on the plants' chemical and structural defenses against herbivores. Chemical defenses include secondary compounds which deter ingestion and/or the normal functioning of the herbivores' digestive system (Rosenthal and Janzen 1979). For instance, terpenes in forage plants can inhibit the activity of the microflora in the rumen of deer (Oh et al. 1968).

The other common kind of defense employed by plants against herbivores is the deposition of indigestible compounds and structures such as lignin, silica, and thorns. These structural defenses not only dilute the nutritionally useful fraction of the diet, but also render the food less digestible (e.g., Demarquilly and Jarrige 1974). Thus leaves are preferred over stems, which have more structural defenses, and grasses are often preferred over aromatic forbs rich in secondary compounds. However, when the available quantities of the preferred components drop—as a result of grazing pressure, or maturation of the sward—then a threshold is reached where the herbivore trades off food quantity against quality and eats less selectively in order to maintain the rate of intake.

Several methods have been used for the determination of the diets of large mammal herbivores: the commonest are direct observation of feeding individuals (the bite-count or feeding-minute methods, Allden and Whittaker 1970, Field 1972); fecal analysis (Holechek et al. 1982a); stomach content analysis (Gwynne and Bell 1968); and esophageal fistulation (van Dyne and Torrell 1964). Though each of these methods is widely used, each has disadvantages when judged according to the criteria of accuracy, representativeness, and cost-effectiveness. For this study, microhistological analysis of fecal material was used, as it has been shown in other studies that the monogastric digestive system of horses has little or no effect on the species composition of the plant material as it passes through the animal (zebra, Stewart 1967, Owaga 1977). This was tested by S.T. Skelton, who fed stabled horses diets of known composition and determined the fecal proportions of the dietary components. The only demonstrable bias among the Camargue plants was that the legume *Trifolium repens* was underrepresented in the fecal proportions (Skelton 1978).

Fecal analysis has been used in many other studies of equids (e.g., Gwynne and Bell 1968; Owaga 1977; Hansen and Clark 1977; Olsen and Hansen 1977). Although zebras and horses are often considered pure grazers, when grass is sparse, they switch to some dicotyledon species (e.g., Hansen 1976). In only one of these studies has the diets of animals of different ages and sexes been examined (Lenarz 1985). In the Camargue study, there were horses of both sexes, a wide range of ages and more than one herd; and the social status and relationships of the horses were known (Chapter 6).

The nutritional value of the diet is more difficult to measure; horses require adequate amounts of energy, as well as a wide range of nutrients, such as nitrogen, minerals, oligo-elements and some vitamins. In practice, free-ranging animals generally obtain adequate amounts of oligo-elements and vitamins from green plant matter. Of the minerals, sodium deficiencies are relatively common and are often made up by the use of salt licks. In the Camargue, high levels of sodium in the soils and in halophytic plants (Bassett 1978) mean that adequate amounts of sodium are always available. Calcium/phosphorus deficiencies and imbalances are more intractable and can cause serious metabolic disorders. The concentrations in plants which are required to provide adequate dietary levels of these elements for horses are known (NRC 1978).

The commonest deficiency in the diet of free-ranging herbivores is a protein/energy deficiency. Levels of digestible energy and protein are closely correlated in natural forages, and protein deficiency causes depressed appetite in equids, and thus inadequate intake of energy (NRC 1978). Dietary crude protein is therefore an index of nutritional value which is widely used. Fecal crude protein is a good measure of dietary crude protein in herbivores which are not eating foods rich in tannins (Bredon et al. 1963, Robbins 1983); this was not the case here, so I have used it in this study (see also Salter and Hudson 1979).

The nutritional value of the diet of herbivores is determined not only by the concentrations of nutrients, but also by both the digestibility of the food and the amount eaten. In this chapter, I first describe the botanical composition and seasonal variations in the diet and next evaluate the causes of variations among individual animals. The nature and extent of feeding selectivity and the quality of the diet are also presented. The digestibility and quantities of food eaten are then estimated, and the amounts of key nutrients eaten are compared with the requirements of domestic horses (Robinson and Slade 1974, NRC 1978). Further information on the methods used is given in Appendix 6.

Diets

Differences among Seasons and among Herds

The horses ate a wide variety of plant species: of the 106 species of vascular plants which have been recorded in the study area (Bassett 1978), frag-

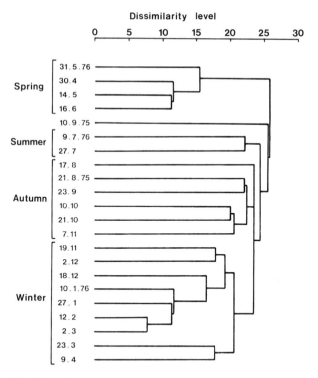

Fig. 1. Classification of the sample weeks by single-linkage cluster analysis. (Redrawn with permission from Skelton 1978.)

ments of 33 species were found in the fecal material. Monocotyledons composed >90% of the material, except in winter, when the halophytes *Halimione, Arthrocnemum*, and other dicotyledons contributed about a third.

The average composition of the material from the older horses (>2 years) for each sample during the period September 1975–August 1976, when samples were taken twice in most months, was calculated and the 22 samples classified using single-linkage cluster analysis; see Fig. 1.

The only distinct cluster is composed of the four samples from the spring (30.4–16.6). Within the rest, there is another relatively well-defined cluster comprising the samples from midwinter (18.12–2.3), which was flanked by two pairs of samples before (19.11–2.12) and after (23.3–9.4).

The samples from the summer and autumn were more heterogeneous; nonetheless, the July samples were distinct from an autumn cluster (21.8–7.11). Apart from samples which were only weakly linked with the others of that season (17.8 and 10.9), the samples clustered according to season, so for the remainder of the results, I have divided the samples into the four seasons shown on Fig. 1. As cluster analysis is a representative and not a statistical technique, the significance of differences among the seasons was tested using an analysis of variance (ANOVA) without replication of a

Table 1. Analysis of Variance on the Botanical Composition of the Fecal Material from Adult Horses

Source	Sum of squares	df	Mean square	F	p
Plant	8,114.1	14	579.6	44.8	0.001
Season	117.0	4	29.2	2.26	n.s.
Herd	14.9	1	14.9	1.54	n.s.
Interaction Plant/Season	8,339.2	56	148.9	11.5	0.001
Interaction Herd/Season	2.3	4	0.57	0.05	n.s.
Interaction Herd/Plant	786.0	14	56.1	4.34	0.001
Interaction H/S/P	723.8	56	12.9		
Total	18,097.3	149			

Note: data from Skelton 1978.

Species × Season matrix. The values in the cells were the average percentage of each plant in the material from the adult horses, transformed by arcsine. Inclusion of a separate matrix for each herd allowed the significance of differences in dietary composition among different herds and seasons to be tested at the same time, using a three-way ANOVA. In this case, the main effects Season and Herd were of no interest because the totals did not vary.

The interaction Plant/Season, Table 1, was highly significant, confirming the validity of the classification. Herd/Season, however, was not significant, showing that the classification, based only on the data from the breeding herd, was valid for the bachelor herd too. Nonetheless, Herd/Plant was highly significant, showing that the diets of animals in the two herds did differ. This point is pursued in the following section.

The seasonal diets of the adult horses are summarized in Table 2, which includes the results for the period September 1976–March 1978, separated into the same seasons. Plant taxa are grouped according to their ecological characteristics into annuals and geophytes, which start growing early in the warm season; marsh edge grasses, which grow later; and perennials, the aerial parts of which remain partly green all through the year.

Both springs were characterized by large proportions of the annuals and geophytes (62, 83%), particularly those characteristic of the wetlands (*Juncus, Scirpus*, and *Phragmites*). In the summers, the wetland plants again dominated the fecal material: though *Juncus* was little used, the marsh edge grasses were important, together composing 11 and 18%. The results for the autumn were rather similar, but perennial grasses were much more important, composing 30–50%. Perennial halophytes were unimportant in any of these seasons, but composed 20–40% of the fecal material in the winters. In the winters, the wetland plants composed only about 10% of the material.

The predominance of the annual and geophyte plants in spring and sum-

Table 2. The Average Botanical Composition of Fecal Material from Adult Horses in the Breeding Herd

Plant	Spring		Summer		Autumn			Winter		
	1976	1977	1976	1977	1975	1976	1977	1975–1976	1976–1977	1977–1978
Annuals and geophytes										
Juncus	**11.8**	**28.0**		1.7			5.1	2.7	9.5	3.7
Bromus	9.8	7.1	0.7	7.2		3.8	1.2		2.3	
Carex + Hordeum	8.0	**12.6**	3.1		0.3	2.0	4.8	1.7	6.7	9.3
Scirpus	**16.1**	**31.1**	**11.3**	**25.9**	**23.8**		**20.5**	3.8	7.4	2.4
Phragmites	17.8	4.2	**36.1**	**11.6**		**10.3**	9.6		1.6	0.5
Marsh grasses										
Paspalum			**16.0**	1.5	**15.0**	9.8	**22.7**	8.9		1.0
Aeluropus	0.8	2.3	1.6	9.3	3.5	6.1		2.0		1.6
Perennial grasses										
Dactylis	**16.8**	6.2	**12.7**		1.3					
Agropyron	**14.9**	3.5	7.7	**22.5**	**30.1**	**37.8**	**20.3**	**40.0**	**19.1**	**34.6**
Brachypodium	0.8		1.4	**11.6**	**16.9**	**11.4**	**9.4**	**10.8**	8.2	7.2
Perennial halophytes										
Halimione	0.2		1.8		7.2	8.1		**20.2**	**18.6**	**33.2**
Arthrocnemum							0.6	1.6	22.8	3.2
Other	3.0	5.0	7.7	8.7	1.9	11.7	5.8	8.3	4.0	3.6

Note: Data from 1975–1976 from Skelton 1978.

mer diets presumably comes from the fact that their aerial parts, being dead in the winter, have mostly become litter by the time the new growth starts in spring. The horses are therefore able to harvest fresh green matter, which is of very high nutritional value (see Chapter II, Fig. 12) and is undiluted by the nutritionally poor dead shoots: The horses did not eat these at all. In addition, *Phragmites* and *Scirpus* are so large that the horses often harvest a single shoot in a bite, and it is relatively easy for the animals to brush aside any remaining dead shoots with their lips. These large marsh plants were the only species in the annual + geophyte group to be used intensively in the summers (see Table 2). The others, the *Juncus* species and annual grasses *Bromus* and *Hordeum*, grow on higher ground and have shorter seasons of vegetative growth: by July, they appear overmature and very fibrous. Spring and summer were also notable for the near absence of the *Arthrocnemum* and *Halimione*, which together formed 20–40% of the winter diet. The marsh-edge grasses *Paspalum* and *Aeluropus* together formed a significant part of the diet in autumn (16–23%) but not at other times of the year; and by the autumn, perennial grasses were forming as large a proportion (30–50%) of the diet as they did in winter. Thus, the main change from autumn to winter was the remarkable increase in the use of the dicotyledonous halophytes *Halimione* and *Arthrocnemum* (from c. 5% to c. 30%).

Arthrocnemum was the only shrub or tree which made up a significant part of the diet in any year. Yet elms (*Ulmus campestris*) and willows (*Salix gracilis*), among the trees, were widespread; among the shrubs, *Phillyrea angustifolia* and *Tamarix gallica* were abundant; and the two species of *Arthrocnemum*, *fruticosum* and *glaucum*, were the dominant plants over much of the area (Bassett 1978). The horses therefore avoided the shrubs and trees, as do equids generally (cf. references in Fig. 11).

The switch from annuals and geophytes in the growing seasons to perennial grasses in the autumn, and to the abundant perennial halophytes in the winter suggested that the horses may have been seeking out the greenest food plants in each season. The nutritional consequences of such selective behavior is examined in the next section, and their responses to spatial variations in the abundance of green plant matter is examined in the chapter on habitat selection (Chapter V).

Longer-Term Trends: Differences among Years

Superimposed on the seasonal changes in diets were considerable differences across years for some taxa. In particular, *Arthrocnemum* was heavily used in winter of 1976–1977, but not in the previous or subsequent winter; see Table 2.

The use of both *Phragmites* and *Paspalum* declined between years in each season; and in parallel fashion, there was an increase in the use of ecologically similar plants: *Scirpus*, which grew in the marshes with *Phrag-*

mites, and *Aeluropus*, which was ecologically similar to *Paspalum*. The horses appeared to eating out *Phragmites* and *Paspalum*, both of which were preferred species in the growing season.

Variations among the Diets of Individual Horses

The diets of the horses were affected by the herds they lived in because there was a significant interaction of Herd/Plant; see Table 1. On a year-round basis, the two herds differed in their use of two species only, the marsh edge grasses *Aeluropus* and *Paspalum*. The rankings for the principal plants in the fecal material of the two herds are shown in Table 3; for

Table 3 Range Test on the Average Botanical Composition of the Fecal Matter of the Adults of the Two Herds

Taxon	Herd	Composition (%)	Significant differences
Agropyron	BR	29	
Agropyron	BA	28	
Phragmites	BR	25	
Phragmites	BA	24	
Aeluropus	BA	19	
Paspalum	BR	18	
Halimione	BA	15	
Brachypodium	BA	14	
Halimione	BR	12	
Brachypodium	BR	12	
Dactylis	BR	11	
Dactylis	BA	11	
Trifolium	BR	11	
Aeluropus	BR	10	
Suaeda	BA	10	
Trifolium	BA	9	
Suaeda	BR	9	
Other	BR	8	
Carex + Hordeum	BR	8	
Carex + Hordeum	BA	7	
Other	BA	6	
Bromus	BR	5	
Juncus	BR	5	
Juncus	BA	4	
Bromus	BA	4	
Paspalum	BA	3	
Scirpus	BA	2	
Scirpus	BR	2	

Note: BR = breeding herd; BA = Bachelor herd; Mean not significantly different are joined by a bar. The only two species for which the means were significantly different ($p < 0.05$) between herds were *Aeluropus* and *Paspalum*; data from Skelton 1978.

Fig. 2. Distribution of the two species of marsh-edge grass in the study area.

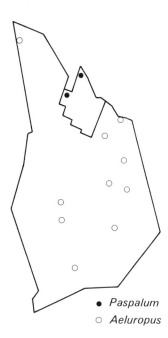

● Paspalum
○ Aeluropus

each herd, one species ranks third and the other much lower. Apart from this difference, the diets were very similar. *Paspalum* was a rare plant, composing about 1% of the sward, and was found in the fallow fields only, while *Aeluropus* was commoner and more widely distributed; see Fig. 2 and the following section on *Selectivity of feeding*."

The home ranges of the herds were not identical during the year 1975–1976, (see Chapter VI, Fig. 1), and it is clear that the bachelor herd rarely used the fields. The difference between herds in the use of *Paspalum* was presumably a consequence of this difference in home ranges. It is striking that the place of *Paspalum* in the diet of the bachelors was taken by the ecologically similar *Aeluropus*. Further, the change occurred immediately when a horse transferred to the bachelor herd; see Table 4.

The causes of the differences among the diets of individual horses were analyzed using correspondance analysis. The main source of variation was the herd to which a horse belonged; see Fig. 3. When eliminating the bachelors from the analysis, it was found that within the breeding herd, the main source of variation was age, or a correlate of age; see Fig. 4. In the spring data, Axis 1 (accounting for 20% of the variation), opposed a pasture grass (*Dactylis*) to a rush (*Scirpus*), which was eaten more by the young horses (see Fig. 4, a).

In the summer data, Axis 1 (accounting for 20% of the variance), opposed the rare marsh edge grass (*Paspalum*, 0.9; see Table 5 to a common perennial grass (*Agropyron*, 6.6%; see Fig. 4, b).

Table 4. Relative Amounts of the Marsh Edge Grasses in the Fecal Matter of the Breeding Herd and the Bachelor Herd

Species	July 9, 1976		July 27, 1976		August 17, 1976	
	Ae	Pa	Ae	Pa	Ae	Pa
Breeding herd	2.9	10.9	0.3	21.1	12.2	19.7
1	3	17	3	11	9	25
J1	4	14				
H1	3	14				
			Herd transfer ↓			
J1			23	0	34	0
H1			19	0	39	0
Bachelor herd	9.2	4.0	18.6	0.0	42.0	0.0

Note: Data gathered at three times in the summer. Also shown are the data for a male which did not change herds ("1") and for two males that did ("H1, J1"; *Aeluropus*, Ae, and *Paspalum*, Pa; data from Skelton 1978).

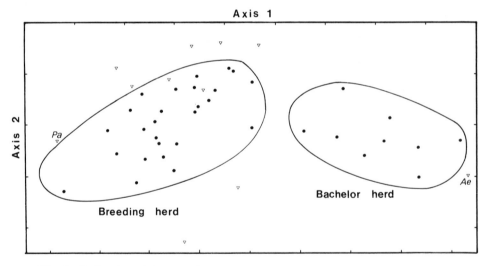

Fig. 3. Ordination of the summer diets of all the horses. Filled circles = individual horses; open triangles = plant species. Only those important in the construction of Axis 1 are given: Ae = *Aeluropus*, Pa = *Paspalum*. (Data from Skelton 1978.)

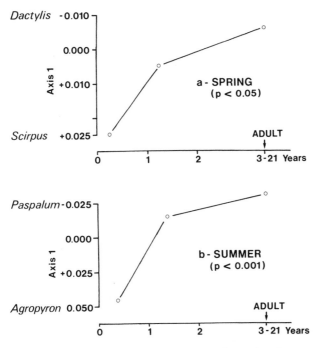

Fig. 4. Positions on the "age axis" of the ordinations (a, spring; b, summer) of the different age classes of horses; the grass species associated with the axes are shown on the left, and the significance of differences between age classes is given on the figures, tested with analysis of variance. (Data from Skelton 1978.)

In the analysis for the autumn, Axis 1 (25%) was uninterpretable. However, Axis 2 (22%) was again related to the age of the horses and opposed the perennial grass *Agropyron* (4.5%) to the very common *Halimione* (38%) (see Fig. 5, a).

In the winter data, Axis 1 (37%) opposed the rare *Paspalum* (0.8%) to the common perennial grass *Brachypodium* (10.7%) (see Fig. 5, b). In the last three analyses (summer, autumn, and winter), the adult horses' diets contained more of the rarer of the plants in each case.

There was no effect of the animals' gender on the plant species composition of the feces. These results confirm, but only in part, those of Lenarz (1985), who found no significant differences in the diets of males, females, and foals. However, rare species were eliminated from the analysis, which could explain why no differences were found between adults and young horses in that study. The fact that most of the species preferred by the Camargue adults were rare implies that these horses developed the ability to locate, or to monopolize them with age. The nutritional consequences of this behavior is evaluated next.

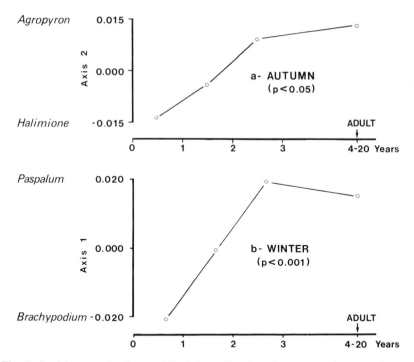

Fig. 5. Positions on the "age axis" of the ordinations (a, autumn; b, winter) of the different age classes of horses; the grass species associated with the axes are shown on the left, and the significance of differences between age classes is given on the figures, tested with analysis of variance. (Data from Skelton 1978.)

Proximity and Diet Similarity

There were therefore consistent differences among individuals' diets: some horses ate more of certain plants (e.g., the marsh edge grass *Paspalum*) than did others. Quantitatively, the greatest differences were accounted for by the herd to which the horse belonged, and when a horse changed herd, it changed its diet. This implies that the differences in diet were due to differences in the availability of the plants in the largely separate home ranges of the herds. However, there remained the possibility that the horses were cuing from each other in their choice of food plants. If such tactics are an important feature of the foraging of horses, then one would expect animals which spent a large proportion of time foraging in close proximity to have diets which resembled each other closely. In this herd, some horses had strong social bonds, expressed by the maintenance of close proximity, kinship groups centered on the mares, and peer groups based on age (Wells and v. Goldschmidt-R. 1979).

As measure of diet similarity, the euclidean distance was calculated between each horse's diet and all the others. Using the winter data, the distance between the diet of each mare and her offspring was then compared with the average distance between the mares and the peers of their offspring. The mean distance between mothers and their offspring (10.7) was not significantly different from the mean distance between the same mares and the peers of their offspring (10.5, $d = 0.17$, $t = 0.54$, $df = 14$, $p < 0.5$).

Within age classes, I examined the relationship between time spent as close neighbors and diet similarity: there was no significant relationship in any of the age classes examined—foals, yearlings, and adults ($r^2 = 0.03$, 0.001, 0.005; $n = 21$, 11, 28, respectively).

Diet similarity therefore was not statistically related to time spent in close proximity either at the family, or at the peer-group level. The horses did not seem to be cuing from each other.

Selectivity of Feeding

The proportions of each taxon in the sward and in the fecal material are shown in Table 5, together with the preference index calculated as described in Appendix 6.1. Because the fecal proportions varied much more than did the sward proportions (e.g., *Agropyron* in the feces 7.7–40% vs. in the sward 4.5–9.0%), the principal plants tended also to be the preferred ones. The horses always fed selectively: in each season, there were plants which made up five times more of the diet than of the sward. Selectivity was less marked in the winter (there were two preferred species) than at other times of the year, especially the spring, when six taxa were preferred. No species was constantly preferred, but the horses did show very strong preferences for *Agropyron* in autumn and winter, and for *Phragmites, Paspalum*, and *Juncus* in the warm seasons.

In spring, the horses showed marked preferences for plants with aerial parts that die in winter (annuals and geophytes), which were thus entirely composed of green tissues in the growing season. In winter, though, they were less selective in their feeding; the horses switched to perennial grasses and forbs, which still had some green parts.

Nutritional Value

Mineral Nutrition

Calcium and phosphorus levels in the principal food plants of the horses are presented in Table 6. Calcium (Ca) levels were high in two of the species in winter; in spring, only three of eight species had Ca levels within the range recommended by NRC (1978) for yearling horses of mature weight (425 kg). Phosphorus (P) levels, on the other hand, were low in all plants,

Table 5. The Average Botanical Composition of the Fecal Matter of Adult Horses in the Breeding Herd, the Composition of the Sward Available and the Preferences Shown by the Horses

Taxon	Spring			Summer			Autumn			Winter		
	Available	Fecal	Preference	Available	Fecal	Preference	Available	Fecal	Preference	Available	Fecal	Preference
Annuals and geophytes												
Juncus	0.8	11.8	14.8*	0.5	0	0	2.7	0	0	1.1	2.7	2.2
Bromus	4.8	9.8	2.0*	3.2	0.7	0.2	4.6	0	0	4.6	0	0
Carex + Hordeum	2.2	8.0	3.6*	0.5	3.1	6.2*	3.4	0.3	0.1	8.9	1.7	0.2
Scirpus	5.7	16.1	2.8*	5.6	11.3	2.0*	5.6	0	0	4.8	0	0
Phragmites	5.6	17.8	3.2*	5.5	36.1	6.6*	5.7	23.8	4.2*	5.2	3.8	0.7
Marsh edge grass												
Paspalum	0.6	0	0	0.9	16.0	17.8*	1.2	15.0	12.5*	0.8	8.9	11.1*
Aeluropus	2.4	0.8	0.3	3.2	1.6	0.5	3.8	3.5	0.9	3.0	2.0	0.7
Perennial grasses												
Dactylis	7.5	16.8	2.2*	3.3	12.7	3.8*	2.9	1.3	0.4	1.7	0.0	0
Agropyron	9.0	14.9	1.7	6.6	7.7	1.2	4.5	30.1	6.7*	5.0	40.0	8.0*
Brachypodium	2.3	0.8	0.3	16.2	1.4	0.1	13.7	16.9	1.2	10.7	10.8	1.0
Perennial halophyte												
Halimione	45.3	0.2	0.0	38.9	1.8	0.0	38.1	7.2	0.2	35.5	20.2	0.6
Other	14.6	3.0		15.4	7.7		15.3	1.9		19.4	8.3	

Note: * Selection significant at $p < 0.05$, chi-square test. From Skelton 1978 with permission.

Table 6. Dry Matter Concentrations (%) of Ca and of P, and the Ratio of the Two

Plant taxa	Element	Month		
		Jan	Feb	May
Halimione	Ca	*0.89*	*0.96*	—
	P	0.11	0.12	—
	Ca/P	*8.1*	*8.0*	—
Brachypodium	Ca	0.24	0.29	0.33
	P	0.06	0.08	0.20
	Ca/P	*4.0*	*3.6*	*1.65*
Agropyron	Ca	*0.65*	*0.79*	*0.85*
	P	0.10	0.11	0.22
	Ca/P	*6.5*	*7.2*	*3.9*
Trifolium	Ca			*1.24*
	P			0.21
	Ca/P			*5.9*
Juncus gerardii	Ca			*0.84*
	P			0.20
	Ca/P			*4.2*
Bromus	Ca			0.26
	P			0.24
	Ca/P			*1.08*
Phragmites	Ca			0.28
	P			0.27
	Ca/P			*1.04*
Scirpus	Ca			*0.29*
	P			0.25
	Ca/P			*1.16*

Note: Values within the recommended range for lactating mares are italic (Ca > 0.47, $p > 0.39$, ratio Ca/P 1.1–5.0; NRC 1978—Table 3). The ranges recommended for mature horses at maintenance are Ca > 0.31, $p > 0.24$, ratio of Ca/P 1.1–5.0.

especially in winter. As a consequence the Ca/P ratio was too high in winter, but about right in spring.

Phosphorus concentrations varied from one part of the plants to another and were especially high in the green shoots of *Phragmites*; see Table 7. Nonetheless, P concentrations equal to or above those recommended for lactating mares were found only in *Phragmites* in March and April. The level recommended for maintenance (0.24%) was found in all species in spring.

The distribution of P in the plants was closely correlated with the distribution of crude protein, both among plant parts and among swards; see Table 7.

These nutrients were highest in the green, growing tissues, and especial-

Table 7. Phosphorus and Crude Protein Concentrations in Various Plant Parts during 1975

Month	Taxon	Part	Phosphorus (%)	Crude protein (%)
March	*Juncus gerardi*	Green leaf	0.35	22.4
	J. gerardi	Brown leaf	0.11	10.9
	J. gerardi	Brown stem	0.11	9.8
	Agropyron	Green leaf	0.30	17.2
	Agropyron	Green stem	0.34	12.5
	Agropyron	Brown leaf	0.09	6.7
	Agropyron	Brown stem	0.05	4.5
	Phragmites	Whole shoot	**0.44**	23.0
April	*Phragmites*	Whole shoot	**0.43**	21.7
May	*Phragmites*	Green shoot	0.33	21.9
	Phragmites	Brown leaf	0.14	8.9
June	*Phragmites*	Green leaf	0.23	17.6
July	*Phragmites*	Green leaf	0.14	13.5
	Phragmites	Brown leaf	0.08	5.2
August	*Phragmites*	Green leaf	0.14	11.1
September	*Phragmites*	Green leaf	0.17	14.9
	Phragmites	Brown leaf	0.07	5.7
October	*Phragmites*	Green leaf	0.17	14.6

Note: The two variables are closely correlated both in this data set ($r = 0.90$, $N = 18$, $p < 0.001$) and for whole swards across vegetation types and months ($r = 0.88$, $N = 72$, $p < 0.001$).

ly in *Phragmites*. By feeding selectively in the growing season on plants with all-green aerial parts, especially *Phragmites*, the horses improved the nutritional quality of their diets.

Protein

In spite of this selective behavior, the quality of the diet varied widely through the year. The lowest values of fecal crude protein occurred in winter (see Fig. 6), and the highest in spring (April), with summer and autumn intermediate. The median values here, and the seasonal pattern, are closely similar to those observed in a study of feral horses in Canada (6–13% CP_f, Salter and Hudson 1979).

Using Equation 6.2.2, given in Appendix 6, approximate values of CP_d, (dietary crude protein) were calculated: these varied from 20% in April 1975 to 4.6% in February 1979. The recommended level for lactating mares of body weight 425 kg (13.3% $\equiv$ 11% CP_f, NRC 1978) was exceeded, in the median year, only in April. Even the recommended level for

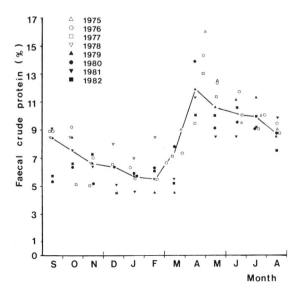

Fig. 6. Seasonal variations in the fecal crude protein of adult mares over 8 years. The line joins the medians for each month.

maintenance ($10\% \equiv 8.5\%$ CP_f) was reached only in the months April–September. The horses' diet in the median year was therefore protein deficient for 6 months.

The longer-term trends in diets (see above), in particular the decline of *Phragmites* and of *Paspalum*, had consequences for the quality. As the years went by, there was a tendency for diet quality to decline in all seasons (see Fig. 6), though the change was significant for spring only (see Fig. 7).

During the year when the bachelor herd had a range that was geographically separate from that of the breeding herd, their diet was of a lower quality; see Fig. 8. However, in the years when the herds' ranges overlapped (1977–1978), there were no significant differences.

In the breeding herd (1979–1982) adult males had lower fecal crude protein concentrations than females; see Fig. 9. The difference was most marked in spring, when the value for males was 20% below that of females.

In 1976, when there were several males and several females in the breeding herd, within each sex, fecal crude protein was correlated with rank (ranks from Wells and von Goldschmidt-Rothschild 1979; $r = -0.55$, -0.76, $n = 16, 8$; $p < 0.05$, 0.05 for females and males, respectively).

Rank and age were almost perfectly correlated ($r = 0.99$, Wells and von Goldschmidt-Rotheschild 1979), so the analysis was carried out on the data for breeding females alone. The result was again significant ($r = -0.73$, $n = 10$, $p = 0.05$), which suggested that this may be an effect of rank rather rather than of experience. The data for breeding females and for all males are shown in Fig. 10.

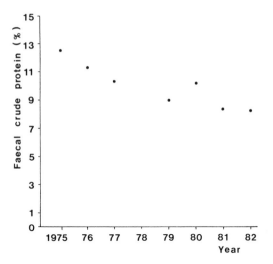

Fig. 7. The trend in fecal crude protein during the spring between 1975 and 1982. The decline is significant ($r_s = 0.96$, $n = 7$, $p < 0.01$); the results for the other seasons also tended to decline but not signicantly ($r_s = -0.16, -0.50, -0.61$, N = 7, 7, 7).

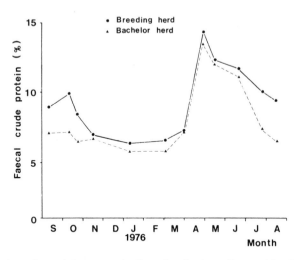

Fig. 8. Fecal crude protein concentrations for the breeding and bachelor herds in 1976. The difference is significant ($d = -1.36$, $t = 4.33$, $df = 10$, $p < 0.01$). The differences between the breeding herd and the bachelor herd disappeared in 1977–1978 ($d = 0.01, 0.26, 0.21$; $t = 0.16, 0.61, 0.55$, $n = 11, 11, 11$).

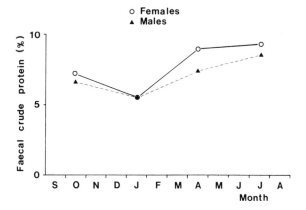

Fig. 9. Mean values for adult female and male horses' fecal crude protein in autumn, winter, spring, and summer of 1979–1982. The difference between the sexes was significant ($d = 0.77$, $t = 2.23$, $df = 15$, $p < 0.05$).

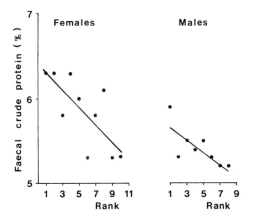

Fig. 10. The relationships between social rank and fecal crude protein concentration for male and female horses, 1976. Both regressions are significant ($r = 0.73$, 0.76; $n = 10$, 8; $p < 0.05$, 0.05) and suggest that the fecal crude protein of dominants is 10–20 % higher than bottom-ranking horses.

I showed in Figs. 4 and 5 that the older horses ate more of certain rare plant species than did the younger animals. This difference in diet may explain the higher quality of the diets of the older, dominant horses. I examine the consequences of this effect in Chapter 7, where the data on growth rates are presented.

In spite of their similarity in botanical terms, the diets of adult male horses were of inferior quality to those of females. This was true not only when the males were in a bachelor herd with a separate home range, but

also when they were in the same groups as the females. This kind of difference between the sexes is widespread in dimorphic ungulates (e.g., red
deer, Clutton-Brock et al. 1982, p. 247) and tends to be most marked in
winter. The usual explanation of this difference is that males, being larger,
require a higher daily intake but can tolerate a lower quality than the
females. This leads them to feed in different vegetation types and to a
degree of segregation of the sexes in winter.

Among monomorphic mammals in which the sexes do not segregate,
male and female klipspringer eat similar foods (Dunbar and Dunbar 1974),
while in indri (*Indri indri*) males had poorer diets than females. This was
ascribed to the fact that they tended to live on the edge of groups and to be
excluded from the best feeding sites by the, dominant, females (Pollock
1977).

The difference between the diets of male and female horses living in the
same herds was imperceptible in winter and tended to be greatest in spring.
It was clearly not a consequence of segregation between the sexes; and
females did not dominate adult males (see Chapter 6). Male horses may
have exercised less feeding selectivity in spring because they directed some
of their feeding time to competition for females. This hypothesis is tested
in Chapter 5.

Conclusions

Because the concentrations of crude protein and phosphorus were closely
correlated, a feeding strategy which maximized one of these nutrients
would therefore maximize the intake of the other, too.

These results show that in spite of the horses' highly selective feeding,
the quality of their food was low: for much of the year, phosphorus and
crude protein occurred at concentrations which were below those recommended for lactating mares. Even for horses at maintenance, the phosphorus concentrations were low for 9 months, and protein was low for
6 months.

In these circumstances, even by feeding selectively, the horses could
lactate and grow only by eating more dry matter per day than would
domestic horses.

The Quantities of Food Eaten

Total fecal collections were made over 24 hr from four mares a total of nine
times between January and June 1979. The results are shown in Table 8,
together with estimates of daily dry matter intake (DMI). The estimates
varied between 3.4 and 4.5% of liveweight (median = 4.2).

Domestic horses weighing 425 kg and eating food of an appropriate
quality consume the equivalent of 1.3–2.2% of their liveweight per day
(NRC 1978). Cattle of the same weight as these horses eat food weighing

Table 8. Estimation of Food Intake by the Mares

Horse	Period	Date	Estimated liveweight (W, kg)	(W$^{0.75}$)	Fecal output (kg$_{FW}$)	Dry matter (%)	Fecal output (kg$_{DM}$)	CP$_f$ (%)	CP$_d$ (%)	DDM (%)	Estimated DMI (kg/d)	DMI (%LW)	DMI (gW$^{0.75}$/day)
I7	24 hr	1/17–18	380	86	41.74	23.6	9.85	5.5	5.9	41.8	16.9	4.5	197
C1	24 hr	2/10–11	420	93	52.54	18.7	9.82	7.15	8.2	45.4	18.0	4.3	194
I2	Day time	3/8–23	350	81	31.29	20.5	8.41	6.5	7.3	44.0	15.0	4.3	185
I7	24 hr	5/9–10	340	79	37.64	15.2	5.72	13.2	16.2	58.1	13.6	4.0	173
	Day time	5/14–18			36.81	15.4	5.67	13.5	16.6	58.8	13.8		
I2	Day time	5/21–24^a	330	77	21.95	(17.2)	(3.78)^a	10.3	12.4	52.1	12.4	3.8	160
	Day time	5/25–30			35.68	16.7	5.96						
D2	Day time	6/6–9	410	91	42.07	16.9	7.11	9.6	11.5	50.7	14.4	3.5	155
	24 hr	6/11–12			40.35	16.5	6.66	10.3	12.4	52.1	13.9		

Note: The estimated dry matter intake (DMI) is derived from

$$DMI = \text{fecal output} \times \frac{100}{100 - DMD}.$$

DMD = digestibility of dry matter, is derived from CP$_d$ as explained in the text; CP$_d$ = crude protein in the diet; CP$_f$ = crude protein in the feces; FW = fresh weight; DM = dry matter; W = liveweight.
^{a} Observation when mare had equine influenza; result ignored.

Table 9. Estimated Phosphorus and Protein Intakes

Mare		DMI (kg/day)	Digestible phosporus (g/kg)	Phosporus intake (g/day)	Crude protein (%)	Digestible crude protein (g/kg)	Digestible crude protein intake (g/day)
I7	Jan	16.9	0.42	7	5.9	27	456
C1	Feb	18.0	0.45	8	8.2	42	756
I2	Mar	15.0	0.44	7	7.3	36	540
I7	May	13.6	1.34	18	16.2	115	1,564
I2	May	12.4	1.20	15	12.4	77	955
D2	Jun	14.4	1.17	17	11.5	69	990

Note: The phosphorus requirement for lactating mares of weight 425 kg (18 g/day, NRC 1978) is met only for one mare in May. The protein requirement (768 g/day) is for all the mares in May. The levels in winter are adequate for mature horses at maintenance (6 g/day digestible P, 280 g/day digestible crude protein).

1.5–3% of their body weight per day and consume less as food quality declines (Van Soest 1982, Fig. 17.9). This was not the case in these horses; they tended to eat more in winter than in spring (see Table 8), though the differences were not significant.

Using these figures for intake, the values for CP_d and phosphorus already given (Tables 6, 7), and digestibilities calculated from the equations in Appendix 6 (the digestibility of P was taken as being the same as for dry matter), the estimated daily intake of P and CP has been calculated; see Table 9. The high levels of dry matter intake meant that the protein intake was above the requirement for maintenance in all months and was adequate for lactation in March, May, and June.

Camargue horses therefore were very selective in their feeding and ate very large quantities of the rather poor forage available. This allowed them to obtain amounts of nutrients adequate for growth and lactation in the growing seasons.

General Conclusions

The horses were essentially graminoid feeders as in other ecosystems, though they ate considerable amounts of dicotyledons in the winters. Food selection by equids contrasts with browsers such as deer, and is very similar to sympatric grazing bovids, (Fig. 11). Nonetheless many studies have shown that grazing bovids generally use dicotyledonous forbs and woody plants to a greater extent than equids (e.g., Hansen and Clark 1977, Olsen and Hansen 1977, Krysl et al. 1984). The, presumably chemical, causes of this difference between these animals have not been determined. However the ability of ruminants to detoxify secondary metabolites is well known (Van Soest 1982, Chapter 3; Crawley 1983). Equids are not known to have

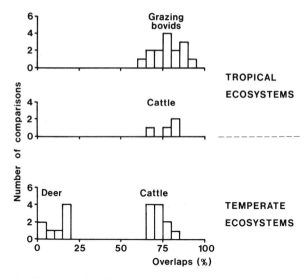

Fig. 11. Overlap between the diets of equids and cattle, and between equids and other grazing bovids. Data for browsing deer, *Odocoileus*, in temperate grassland are included. (Overlap measured Kulcynski's formula in Hansen and Clark 1977; other references include Casebeer and Koss 1970, Gwynne and Bell 1968, Krysl et al. 1984, Olsen and Hansen 1977, Owaga 1975, Stewart and Stewart 1970.)

similar detoxification abilities: further work is required on this issue, but this difference allows bovids to use a range of plant tissues not available to equids, and very likely contributes to the maintenance of higher densities of bovids in most natural ecosystems.

The feeding strategy of the horses had two essential components:

• A high degree of selectivity, accentuated when food was abundant in the warm seasons, and
• The ability to process large quantities of food per day not only in feeding trials with cut hay (Chapter 3) but also when free-ranging.

Diet quality nonetheless showed strong seasonal variations; within these, dominant horses had higher quality diets than subordinates. The effects of these variations on reproduction and growth will be explored in Chapter 7.

The animals' extreme selectivity led to heavy grazing pressure on the preferred species, especially *Phragmites*. This led to a gradual decline in the quality of the diet as the horse population increased: the impact on the plants is examined in Chapter 8.

For a large mammal, consuming large amounts of food while maintaining a high level of selectivity on low quality swards is likely to pose problems of time budgeting. This is examined in the next chapter (5).

Chapter 5   Foraging Behavior

Still out of hardship bred
Spirits of power and beauty and delight
Have ever on such frugal pasture fed
And loved to course with tempests through the night.
                    Roy Campbell
                    from "Horses on the Camargue"

Equids are very effective at extracting nutrients from forages, largely because of their high intake rates (Chapter 3). Most of the studies upon which this conclusion is based were carried out in the stable or laboratory, where the animals were provided with as much hay as they could eat. Free-ranging animals must of course harvest their food for themselves from the forage available.

In Chapter 4, I showed that these horses could extract nutrients at acceptable rates from their coarse forage only by feeding selectively and by consuming large quantities of food (about 16 kgDM per day). If they eat less than their potential maximum intake, which is determined by the volume of their GI tracts and the passage rate of the food, then the daily nutrient extraction rate falls. If the quality of the diet declines (i.e., becomes more fibrous), then again the extraction rate declines sharply (see Chapter 3, Fig. 14).

In the Camargue swards, like most natural ones, high-quality, low-fiber components such as young green leaves are diluted in large amounts of low-quality tissues—overmature leaves and stems (see Chapter 2, "Vegetation".) The horses therefore need to feed selectively on these high-quality components, especially *Phragmites*, while maintaining high rates of food intake.

Locating and ingesting small packets of high-quality items can reduce the rate of food intake in large mammals because only a relatively small weight of food is harvested per bite. There is almost always a trade-off between improving the quality of the diet by grazing selectively and maintaining the rate of food intake (Chacon and Stobbs 1976, and for theoretical treatments, see Belovsky 1986, Illius and Gordon 1987, 1990). Even a small animal such as a sheep compromises: on an Australian sward, Hamilton et al. (1973) showed that sheep fed selectively when there was abundant plant matter, but when there was less than c. 55 g/m^2, they fed unselectively to maintain their intake rates. Comparable studies have not been conducted on grazing in horses, but they can be expected to respond in qualitatively similar ways. Their high daily food intake (see Chapter 3, Fig. 6) may lead them to adopt a less selective feeding strategy earlier than the grazing ruminants.

The daily food intake (g/day) of a grazing animals depends on

- Bite size (g/bite)
- Bite rate (bites/hr)
- Time spent foraging (hr/day)

For a given set of acceptable food items (green leaves etc.), *bite size* will be determined by the size of the animal's incisor row and the size of the items. Bite size in a given species is therefore usually rather constant.

The other parameters, *bite rate* and *foraging time*, however, can vary considerably. Animals can contribute to high rates of intake of preferred food items by modifying their foraging behavior in these two principal ways. First, they can choose to feed preferentially in those parts of their range where the preferred food items are most abundant, thus maintaining a high bite rate. Secondly they can maximize daily foraging time.

In practice, animals may not make these behavioral contributions to their nutrition. If a nutritionally optimal strategy is suboptimal in relation to other biological processes which affect reproductive success, such as predation or competition for mates, then an animal may make a compromise which reflects its own assessment of the costs and benefits of alternative behavioral acts (Sibly and McFarland 1976). Compromises of this sort are well documented in grazing animals. Daily time spent foraging in Soay sheep is normally about the same in males and females: during the rut, the males reduce their feeding time to about a quarter of that of females. This behavioral modification presumably increases their chances of mating successfully, but the decline in feeding time means that they enter winter in poor condition. This is presumed to be one of the causes of the high rate of winter mortality in adult males (Grubb and Jewell 1974).

The relatively high rate of food intake in equids (Chapters 3 and 4) means that the behavior of equids should be particularly closely attuned to nutritional needs. In particular, the following two aspects of foraging should be organized so as to maximize nutrient extraction:

1. The daily duration of foraging should be long and should be less sensitive to competing strategies (e.g., antipredator and reproductive) than in ruminants.
2. Feeding should be concentrated in the habitats where the best-quality food items (green herbaceous plants) are most abundant so as to allow a high bite rate.

There is one further way in which behavior can affect rates of nutrient assimilation: by maintaining a high level of gut fill. An animal which allows the quantity of digesta in its GI tract to decline by not feeding for long periods will suffer a drop in food assimilation. This is likely to be particularly true for equids whose retention times are so short (t_{50} = c. 30 hr; see Chapter 3, Fig. 10) that a 12-hr night without feeding would leave their GI tracts $\frac{1}{3}$ empty. It can therefore be expected that the temporal organization of foraging in equids will be such that a high level of gut fill is maintained.

These three predictions can be tested on the data from the Camargue study and on some comparative information on other equids. Data covering the 24 hr are essential because animals can compensate at night for events in the daytime, such as disturbance by biting flies (Duncan 1985). Details of the methods used can be found in Boy and Duncan (1979), Duncan (1983), the data on foraging behavior were collected by Martin Leverton in 1976–7, see Mayes and Duncan (1986).

Daily Foraging Time

Developmental Changes

A major part of the nutrients required by foals is supplied by their mothers as milk, Fig. 1. As a consequence, foals initially spend much less time feeding than do other horses. The exact amount of time spent foraging by young foals is variable (see Fig. 2) and is clearly affected by the nutritional status of the mother. Thinner mothers presumably produce less milk, and their foals compensated by increasing their foraging time by c. 100%. Similarly flexible behavior has been described in calves (LeNeindre and Petit 1975).

As the foals grow in weight, their requirements increase. This is not matched by their mothers' milk production; see Fig. 1. As a consequence, their foraging time increases progressively to reach the same level as subadults and mares at weaning; see Fig. 3. A similar pattern is observed in cattle, where weanlings may feed for even longer than other categories of cattle (Hodgson and Wilkinson 1967).

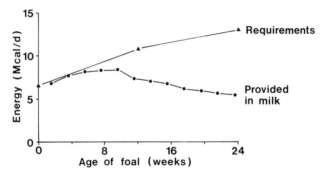

Fig. 1. The mean daily production of gross energy in the milk of mares (data from Bouwman and Van der Schee 1978), and the daily energy requirement of Camargue foals at different ages (calculated using equations in NRC, 1978, and the data for growth of Camargue foals in Chapter VII).

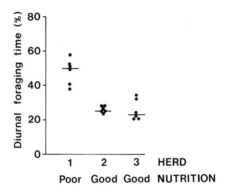

Fig. 2. Daytime foraging time of foals (mean over Weeks 2–11) in three separate herds of Camargue horses. The nutrition of Herd 1 was poor—the mothers lost weight during lactation. In the other herds, nutrition was good, and the mothers maintained or gained weight ($r = 2, p < 0.05$ for Herd 1 vs. Herd 2 or 3; runs test, Siegel 1956; data from Monard 1983).

Reproductive Status

The foraging time of Camargue mares increased at peak lactation and just before weaning, but only by some 8% of their foraging time in the dry period (Duncan 1985). A similar effect was shown in quarter-horse type mares in the United States: lactating females fed for 9.8% longer than nonlactating ones in the same herd (Rittenhouse et al. 1982).

Male horses feed for the same length of time as females outside the breeding season when competition for matings is less strong (October–February, Duncan 1985), or when there are no competitors at all (see Table 1; Kownacki et al. 1978). However, in wild Plains zebra, the males

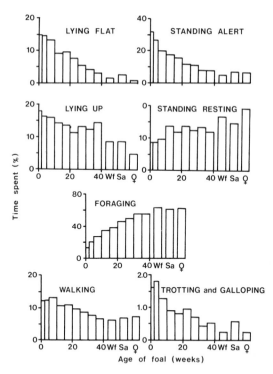

Fig. 3. Changes with age in the time spent in each activity by the foals. For comparison, the times spent by weaned foals (Wf), subadults (the age class 1–2 years older than the foals, Sa), and lactating mares (♀) are shown. (From Boy and Duncan 1979, courtesy of E. J. Brill.)

Table 1. Foraging Times (% in 24 Hrs) in Some Free-Ranging Male and Female Equids

Species	Females	Males	Season	Statistical significance[a]	Source
Tarpan	69.6	70.1	February–September	n.s.	Kownacki et al. 1982
Plains zebra	69.4	58.3	Dry	*	Gakahu 1982
Plains zebra	57.3	55.8	Wet	n.s.	Gakahu 1982

[a]Calculated by source(s)
*$p < 0.01$.

Table 2. Average Time Budgets of Camargue Horses in the Seasons of 1975 and 1976

1975

	Spring		Summer		Autumn		Winter	
	♀	y	♀	y	♀	y	♀	y
Lying flat	1.8a	4.1a	0.7b	1.7b	0.5b	2.3b	0.3b	1.0b
Lying up	9.3a	14.4a	3.5b	8.2b	3.7b	8.6b	4.0b	7.6b
Standing resting	11.7a	7.1a	19.7b	16.2b	21.2b	14.8b	20.7b	16.9b
Standing alert	6.2a	6.3a	8.5b	9.1b	5.8a	5.3a	6.1a	5.9a
Foraging	63.6a	60.6a	56.4b	53.6b	62.7a	62.8a	63.4a	63.0a
Walking	7.2a	7.0a	10.6b	10.0b	5.8a	5.6a	5.3a	5.4a
Trotting	0.1a	0.3a	0.4a	0.6a	0.2a	0.3a	0.1a	0.1a
Galloping	0.1a	0.2a	0.2a	0.3a	0.1a	0.2a	0.0a	0.1a
Rolling	0.0a	0.1a	0.2a	0.2a	0.1a	0.1a	0.0a	0.0a
Number of horses	7	3	7	3	7	3	7	3
Number of days	8	8	10	10	14	14	12	12

1976

	Spring			Summer			Autumn			Winter		
	♀	♂	y	♀	♂	y	♀	♂	y	♀	♂	y
Lying flat	0.6	2.3	1.5	0.4	1.6	2.4	0.6	1.8	3.0	0.2	1.2	1.1
Lying up	7.9	8.5	11.8	2.8	4.2	5.6	3.5	5.3	7.8	3.3	5.8	5.9
Standing resting	16.9	14.3	12.5	20.1	15.5	15.4	22.1	15.1	14.2	22.3	17.1	19.3
Standing alert	7.4	10.8	6.5	8.9	13.8	10.2	5.5	7.1	4.7	5.5	7.0	5.1
Foraging	60.1	55.3	60.0	55.3	51.0	52.1	59.7	61.1	61.5	63.5	64.2	63.4
Walking	6.5	7.2	7.0	12.2	13.0	13.4	7.7	8.7	7.8	5.0	4.3	5.0
Trotting	0.3	0.9	0.3	0.3	0.4	0.6	0.4	0.4	0.6	0.2	0.2	0.2
Galloping	0.2	0.6	0.3	0.1	0.3	0.0	0.4	0.3	0.4	0.0	0.2	0.1
Rolling	0.0	0.0	0.1	0.1	0.3	0.2	0.1	0.1	0.0	0.0	0.0	0.1
Number of horses	9	4	7	9	4	7	9	4	7	9	4	7
Number of days	4	4	8	4	4	4	4	4	4	4	4	4

Note: ♀ = adult females (>3 years or with foals); ♂ = males (>3 years); y = yearlings (12–24 months); significant differences within age classes, among seasons, are shown by different letters (1975 only, paired-comparison t-tests, $p < 0.05$) (from Duncan 1985, courtesy of E.J. Brill).

foraged for less time than did the females, at least in the dry season; see Table 1.

The foraging time of Camargue stallions fell to 92% of the value for females during the breeding season in spring and summer; see Table 2. Interestingly, it was at this time of the year that the quality of the diet of males dropped relative to females (see Chapter 4, Fig. 9), which supports the argument that the very long feeding time observed in these horses was necessary to achieve a high degree of selectivity.

Male and female horses therefore do vary their feeding time, but only slightly: the 24-hr feeding time of weaned Camargue horses varied by less than 10% in relation to age, gender, and reproductive state. The few other data available on the time budgets of the different sex/age classes over 24 hr suggest that this conclusion may be true for equids generally.

Ruminants are much more flexible. In female red deer, lactating hinds feed for 119% of the 9.8 hr grazed by dry hinds (Clutton-Brock et al. 1982), and males may feed for only 25–75% of their normal foraging time in the mating season (e.g., waterbuck, Spinage 1968; Soay sheep, Grubb and Jewell 1974; impala, Jarman and Jarman 1973; blackbuck, Cary 1976; and red deer Clutton-Brock et al. 1982, Fig. 6.21). Insect harassment can cause much greater modifications of feeding behavior in a ruminant, the reindeer, whose daily feeding times may decline by 35% (Skogland 1984, Table 13).

Season

There are also environmentally caused variations in time budgets. At the end of winter, the feeding time of Camargue horses increased to 66% and then fell in June to 51%, Fig. 4.

These variations in foraging time were not significantly related to measures of food quality or abundance (Duncan 1985). The summer decrease was correlated not with temperature but with the numbers of biting flies of the species with long periods of diurnal activity (Tabanidae, Ceratopogonidae, see Chapter 2, p. 48).

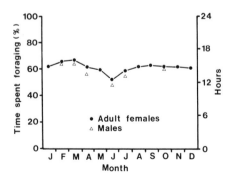

Fig. 4. The mean time spent foraging in each month of the year. (From Duncan 1985, courtesy of E. J. Brill.)

Mosquitos, which have short periods of activity at dawn and dusk (Chapter 2, p. 46), seem to have no effect on 24-hr time budgets, in spite of the obvious distress they cause at these times. Foraging time in the autumn–winter period was correlated with none of the environmental variables tested (humidity, wind speed, rainfall, temperature, biting flies, or food abundance or quality), except sunlight hours, and that was significant in autumn only (Duncan 1985). The slight increase in feeding time at the end of winter occurred when the animals were nutritionally stressed (see Chapter 7) and when their best (green) food was sparsest, so it is possible that they increased their feeding time in order to maximize their intake of high-quality food. However, the increase, c. 10%, was again quantitatively small. Average seasonal time budgets are given in Table 2.

The comparable studies of equids elsewhere, which provide data on 24-hr time budgets, suggest that the results from the Camargue herd are not atypical. Rittenhouse et al. (1982) measured the feeding times of free-ranging cows and quarter-horse-type mares in the United States, using vibracorder devices. This method overestimates the time spent feeding by c. 15% (Ruckebusch and Bueno 1973, Table 1): To allow comparison with the other studies, the data have been corrected accordingly; see Fig. 5. Time spent foraging by the mares was significantly lower in June than in the other months, which averaged 63%, and was much higher than the 40% for the cows.

"Tarpan" (konik) mares in Poland fed for 69% (Fig. 6, Ref. 1); adult Plains zebras in Amboseli, Kenya, fed 59% in the dry and 56% in the wet seasons (Fig. 6, Ref. 5), and in the Serengeti, the zebras fed 62% in the wet season (Fig. 6, Ref. 4). Free-ranging horses and zebras therefore feed for

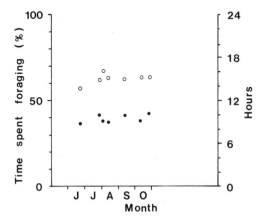

Fig. 5. Comparison of foraging times in mares (open circles) and cows (filled circles) on the same range. (Rittenhouse et al. 1982, modified as described on page 105, with permission.)

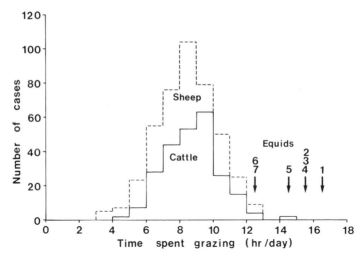

Fig. 6. Comparison of time spent grazing by sheep and cattle (Arnold and Dud-zinski 1978, courtesy of Elsevier) with time spent by equids: (1) tarpan mares (Kownacki et al. 1978); (2) Camargue mares, the present study; (3) range mares (Rittenhouse et al. 1982); (4) zebras, all (Gogan 1973) (5) zebras, all (Gakahu 1982); (6, 7) mares in paddocks. (Arnold and Dudzinski 1978, Doreau et al. 1980.)

similar amounts of the day and for much longer than sheep and cattle (Fig. 6).

When in paddocks, however, domestic horses feed for rather less than the free-ranging equids, 48 and 59% (summer and winter) and 51% (Fig. 6, Refs. 6, 7). The feeding times of stabled domestic horses can be as low as 23% when feeding on concentrates (Ruckebusch 1976).

On Equid Foraging Times

The feeding times of equids are longer than bovids and are clearly related to the ease with which they can obtain their food. Undisturbed free-ranging equids feed for a rather constant 59–69% (14–$16\frac{1}{2}$ hr) per day. Domestic horses feed for less time when on improved pastures, and for very much less when feeding on concentrates. The prediction that the foraging times of free-ranging equids are long and relatively insensitive to competing strategies is therefore upheld by the data.

The upper limit to feeding time appears to be set by the need to perform other essential activities, which are resting, moving, and vigilance (alert behavior). The last two each occupy less than 2 hr per day in winter (c. 6%, Table 2) and include traveling between feeding, resting, and drinking sites, comfort behavior; social behavior; and alarms. It seems reasonable to argue that in most natural habitats, there is little potential for decreasing the allocation of time to these activities without incurring increased rates

Table 3. Matrix of Correlations between the Activities

	LF	LU	SR	SA	F	W
Lying flat (LF)	1					
Lying up (LU)	0.79**	1				
Standing resting (SR)	−0.45*	−0.34	1			
Standing alert (SA)	0.45*	0.34	−0.47*	1		
Foraging (F)	−0.04	−0.26	−0.72**	−0.05	1	
Walking (W)	0.03	0.13	−0.22	0.38	−0.20	1

Note: November–March, $N = 20$ (from Duncan 1985 courtesy of E.J. Brill).
$*p < 0.$; $**p < 0.$

of predation, decreased social cohesion, and less effective use of the food and water resources available. In Camargue horses during winter, time spent walking or alert is unrelated to variations in daily foraging time; see Table 3.

Resting, however, occupied more of the time (23–26%, 6 hrs, in adult females regardless of season or year, Table 2), and it was from resting that extra time for grazing came in winter (Table 3; $r = -0.72$, $p < 0.01$). Assuming that extra feeding time would allow the horses to obtain a better quality diet by feeding more selectively, why did they not take more of the resting time for feeding?

The functions of sleep are not well understood. A widely held view, the immobilization hypothesis (Meddis 1975), is that sleep exists to preserve the inactivity period, increasing the efficiency of rest–activity cycles, but bringing no unique physiological benefit. If this hypothesis is correct, then animals should often show no sleep when there are benefits to using their time for activity.

In terrestrial mammals, though time spent sleeping declines as live-weight increases (Allison and Cicchetti 1976), even the largest animals still invest time in sleep. The shortest times[a] have been reported in captive, tame elephants and giraffes (c. 4 hr, references in Meddis 1975). Undisturbed, free-living elephants rest for 4.8 hr (c. 20%) of the 24 hr (Hendrichs and Hendrichs 1971). Horses, in spite of their relatively small body size, have among the lowest sleep times (3.8 hr, Ruckebusch 1972). These results and the extensive studies of the costs of sleep deprivation (Fishbein and Gutwein 1977) suggest that there are unique benefits to sleep. Four hours per day (17%) may therefore represent an incompressible minimum for mammals.

In the Camargue herd, mares were nutritionally stressed at the end of

---

[a] With the exception of an unconfirmed preliminary report of two shrews, which did not sleep at all (van Twyver and Allison 1969).

winter, as their requirements were high due to pregnancy and/or lactation, and they lost condition. Green plant matter was very sparse, so increasing their feeding time by 4 hr per day could have allowed them, by increasing their feeding selectivity, to improve the quality of their diet while maintaining their daily intake of food. That they increased their feeding time by only 6% implies that the nutritional benefits of further increases in feeding time would be outweighed by the costs of sleep deprivation, or fatigue (Tribe 1950).

This argument can also be used to account for the constancy of equid feeding times. If the benefits of sleeping and moving are relatively constant across seasons, through the year, then the time available for feeding should also remain constant. The feeding times of ruminants appear more sensitive to competing strategies because the duration of foraging is so much more variable than the 10% reported here for equids; see Fig. 6.

Maximizing feeding time within these constraints would allow free-ranging horses to feed as selectively as possible and thus to achieve high rates of food assimilation. The shorter feeding times of horses with low nutrient requirements (adult nonbreeders) and those with superabundant supplies of high-quality food, in stables or in managed pastures, are consistent with this explanation.

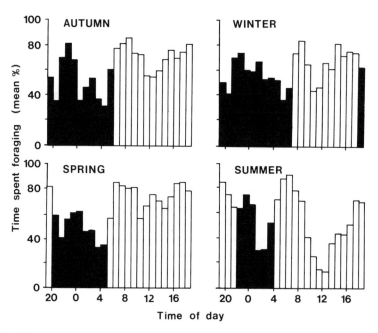

Fig. 7. The average time spent foraging by Camargue mares in each hour; nocturnal hours are shaded. Based on 6 days in each of four seasons in 1975: (a) autumn (7.10–7.11), (b) winter (2.12–11.1), (c) spring (26.3–25.4), (d) summer (9.6–3.7). (From Mayes and Duncan 1986, courtesy of E. J. Brill.)

The Temporal Organization of Foraging Behavior

Feeding at Night

Food harvested by day has higher levels of soluble carbohydrates, by as much as 30% in leaves (Greenfield and Smith 1974), and it is safe to assume that visual cues such as greenness, which horses perceive (Grzimek 1952) and which are presumably used in the selection of food, are more visible by day than by night. Many bovids show such a strong preference for feeding by day that they feed virtually not at all by night (Arnold and Dudzinski 1978).

By contrast, horses and zebras may feed at any time (see Figs. 7, 8), but they usually feed most intensively at dawn and in the afternoon. A rather

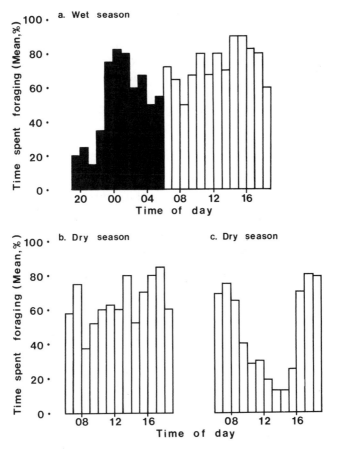

Fig. 8. The average time spent foraging by Serengeti zebra of all ages and both genders in each hour: (a) wet season—January–February, short grass plains; (b) dry season—September, northern grasslands; (c) dry season—July, central woodlands. (Data from Gogan 1973.)

Table 4. The Ratio of Time Spent Foraging by Day (%) and by Night

Species	Season	Day ÷ night	Reference
Plains zebra	Wet	1.0	Gakahu, 1982
(Amboseli)	Dry	1.2	
(Serengeti)	Wet	1.3[a]	Gogan, 1973
Horses	Autumn	0.9	Duncan, 1985
(Camargue)	Winter	1.3[a]	
	Spring	1.5[a]	
	Summer	0.8[a]	

[a] Significant preferences for the day (sunrise to sunset) were found in the zebra in the Serengeti, and in Camargue horses during the winter and spring; Camargue horses preferred feeding by night in summer.

different pattern is evident in the Camargue during the summer (see Fig. 7, d), when the horses rarely fed around midday, apparently to avoid biting flies, which were most numerous and active on the feeding areas (Duncan and Cowtan 1980, Hughes et al. 1981). A strikingly similar pattern was reported when zebras feed in the tsetse-infested central woodlands of the Serengeti; see Fig. 8, c.

In all seasons, and as predicted, equids feed for about as much of the night as the day; see Table 4.

The Organization of Feeding into Meals

I now examine the prediction that the horses organize their foraging in time so as to maintain a high level of gut fill. Foraging in horses occurs in bouts of uninterrupted feeding separated by nonfeeding intervals. Feeding does not occur randomly through the day, but is grouped into "meals" (Mayes and Duncan 1986). The frequency distribution of interval lengths is made up of two populations, short ones of less than c. 10 min, which occur within meals (alertness, walking, etc.), and longer intervals, which separate meals. These longer intervals almost always contain resting and usually sleep (Mayes and Duncan 1986).

Sleep cycles are stereotyped: they have a typical length of 30–60 min (Ruckebusch et al. 1970, Dallaire 1974), and a structure of this kind:

. . . wakefulness–somnolence–quiet sleep[b]–active sleep[c]–wakefulness . . .

It is probable that nonfeeding intervals have a nonrandom distribution of lengths because of inflexibility in the sleep cycle, and it is this pattern which structures feeding behavior in the horse.

---

[b] Quiet sleep = slow waves sleep.
[c] Active = rapid eye movement (REM) sleep.

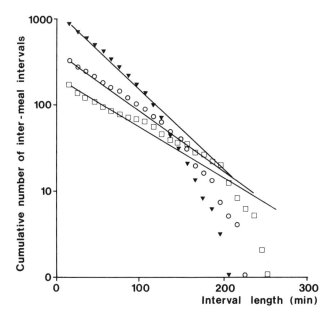

Fig. 9. Log survivorship curves of intervals between meals of Camargue horses during three periods of the year: October–April = filled triangles; May, July–September = open circles; June = open squares. The expected lines, from the negative exponential (random) model, are also shown. (From Mayes and Duncan 1986, courtesy of E. J. Brill.)

Intervals Between Meals

During the cool season in the Camargue, the median length of the intervals between meals was 45 min, and 90% were shorter than 2 hr. The frequency distribution of these intervals (see Fig. 9) was nonrandom, in that there was a deficit of long intervals (Kolmogorov-Smirnov test, $D = 0.075$, $n = 902$, $p < 0.02$); none was longer than $3\frac{1}{2}$ hours.

The horses therefore avoided having very long intervals between successive meals: this pattern of behavior, which contributes to maintaining a high level of gut fill, is found in other species with fast throughput times, such as the roe deer (Hofmann 1989); the physiological mechanisms involved in the control of feeding behavior in stalled ponies have been described by Ralston (1984).

The Effect of Interval Length on Meal Length

Meal lengths are strongly affected by time of day, the longest meals occurring at the daily feeding peaks (Mayes and Duncan 1986). Within periods of homogeneous meal lengths, long intervals tend to be followed by long meals; see Table 5. This preprandial correlation is evidence that once a

Table 5. Correlations between Meal (M) Length and the Length of Previous (Pre-prandial) and Subsequent Intervals (I, Postprandial)

Month	Time	Preprandial	Postprandial
October	19–03.00	0.632**	0.051
	3–05.00	0.679**	0.093
	5–11.00	0.301	−0.046
	11–15.00	0.452**	0.028
	15–17.00	0.089	0.518
	17–19.00	0.385	−0.443
December–	19–11.00	0.278**	0.154
January	11–19.00	0.351**	0.252
April	19–01.00	0.438**	−0.108
	1–05.00	0.206	0.170
	5–07.00	0.359*	−0.584
	7–11.00	0.096	0.131
	11–17.00	0.237	−0.141
	17–19.00	0.762*	0.130
June	19–23.00	−0.338	−0.227
	23–03.00	0.289	0.439**
	3–07.00	−0.161	−0.027
	7–13.00	−0.524**	0.003
	13–17.00	0.261	0.237
	17–19.00	0.481	0.096

Source: Data from Mayes and Duncan 1986, Courtesy of E.J. Brill.
$*p < 0.05$; $**p < 0.01$.

horse begins to feed, it tends to stop at a fixed satiety threshold. In cattle feeding on roughages such a threshold is reached when the gut is full, (Campling 1970, Metz 1975), a homeostatic mechanism which leads to the maintenance of a high average level of gut fill. Experimental work on horses has shown that as the quality of food declines, intake increases, but only to a limited extent (Laut et al. 1985). Their inability to balance their energy intake on very poor foods shows that intake in horses, too, is ultimately limited by gut fill. On the coarse forages of the Camargue, it is therefore probable that their satiety threshold is reached when the GI tract is full.

The Effects of Diet on Meal Length

Meal length in Camargue horses showed no overall seasonal variations, outside the period of fly infestation ("fly period"). In spite of the considerable variations in the abundance of the horses' food, there were no consistent variations in meal length among vegetation types (Mayes and Duncan 1986). Within most vegetation types, there were no significant seasonal variations which could not be explained by the effect of the biting flies. In the deep marsh however, the horses' mean meal length dropped from 210 min in March to 70 min in October. In the Marsh, the horses fed on the green shoots of emergent reeds, the abundance and chemical composition of which could be measured accurately. Meal length tended to decline as the biomass of reeds increased, but the relationship was not significant. Meal length was, however very closely related to the fibrousness of the reeds ($ADF = 32 - 43\%$, $r = -0.99$, $p = 0.01$, $n = 4$). Fermentation rates in the rumen are known to be lower on fibrous forages (cf. Van Soest 1982), so this result implies that the horses were feeding to fill the GI tract, and that this took longer on highly fermentable forages. Similar digestive constraints on maintenance activities of animals have been found in a wide variety of animals, including humming birds (Krebs and Harvey 1986).

Within-Meal Activities

Once feeding started Camargue horses fed in a very stereotyped manner. Their rate of biting remained constant over meals lasting 4–7 hr, which suggests that the motivation of the horse (and observer) did not wane. Similarly, the frequency and duration of standing (alert) and moving intervals showed no consistent change, whether through a meal or across vegetation types (Mayes and Duncan 1986).

Again, the main environmental factor affecting feeding was the rate of attacks by biting flies: when the flies were abundant, the horses frequently stopped feeding and walked forward to rub themselves on another horse or a plant, or stood grooming themselves. As a consequence the mean duration of feeding bouts declined to about 25 sec (see Fig. 10) in dryland vegetation types and about 45 sec in the Deep Marsh.

Conclusions

The organization of foraging behavior in equids is consistent with maintaining a high average level of gut fill:

1. Though generally preferring the day, they also feed at night and avoid long intervals between meals.
2. They feed to a satiety threshold, which is probably reached when the gut is full.

These mechanisms which maintain a high level of gut fill are consistent through the year, except during the 2-month-long fly period, when total feeding time, meals, and feeding bouts are shorter; intervals between meals are longer; and the horses no longer feed to a satiety threshold.

Attacks by the biting flies, especially the Tabanids, therefore modify the temporal organization of foraging of horses in the Camargue to a much greater extent than seasonal or spatial variations in their food supply.

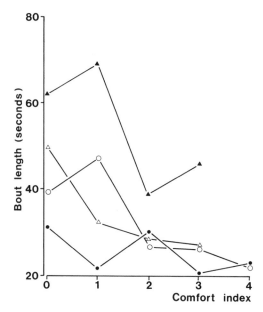

Fig. 10. The relationship between mean bout length and the activity of biting insects. Comfort behavior is used as an index of insect activity (Chapter 2, Fig. 21).
Facets: filled triangles = deep marsh; open triangles = fallow fields; filled circles =
halophyte grassland; open circles = enganes. Comfort behavior index: 0 = none, 1
= infrequent, 2 = frequent, 3 = very frequent, 4 = with rolling and distressed
circling behavior.

Selection of Feeding Habitat by Equids

In this section, I examine the hypothesis that horses and other equids select
habitats principally on the basis of the abundance of high-quality food
items. Unlike some herbivores of the desert, such as oryx and camels,
equids do not have particular physiological adaptations for conserving
water (Maloiy et al. 1978, Joubert and Louw 1977). They therefore need
to drink regularly unless the water content of their food is exceptionally
high; this requirement could be a major constraint on their selection of
habitat.

Water

Many studies have shown that the distance from water is a prime determinant of the use of habitat by free-ranging equids. In the Amboseli region,
Plains zebra are widely dispersed in the wet season, but in the dry season,
few are found more than 10 km from water; see Fig. 11. In the Kruger
National Park, zebras moved, on average, no more than $7\frac{1}{2}$ km from water
during the cool season, and even less in the hot season (Young, in Penzhorn 1982b). Similar patterns have been reported for feral asses and

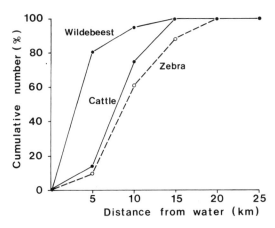

Fig. 11. Cumulative numbers of zebras and of the two commonest grazing bovids, as a function of distance from surface water in the Amboseli region during the dry season. (Data from Western 1975.)

horses in the western United States (Berger 1977, Hanley and Brady 1977). The maximum distance equids have been reported to move from water in dry periods is about 20 km (data in Western 1975, Coetzee 1969).

Some grazing bovids such as Oryx can live without drinking, but the species of the tropical savannas, though they have large foraging radiuses (Pennycuick 1979) are no less dependent on water than zebras, Fig. 11. Eighty percent of wildebeest are found <5 km from water, but only 10% of the zebra are.

Habitat Structure

In a detailed study of habitat use by the large mammal herbivores of Transvaal lowveld, where water was not an important determinant of habitat use, zebra densities were significantly correlated with a number of vegetational characteristics of the habitats. They avoided woodlands and preferred habitats where grasses were abundant; they avoided tall grasses and preferred short grasslands with patches of bare soil (Hirst 1975). However, the structure of the habitats did not account for much of the variation in the densities of zebra and other ungulates: other variables must be more important.

Forage Quality

After rain, when water sources are widespread, zebras, like many other grazing mammals in the tropics, disperse away from permanent water, even though forage is apparently plentiful there. They move into arid or semiarid areas (the Namib desert, Joubert 1972; the south eastern plains of the Serengeti, Maddock 1979) or onto ridge tops (Bell 1970, Penzhorn

1982a) where the growth of grass is ephemeral. This dispersion could be a consequence of differences in quality between grasses with short growing seasons and those with long. It is well known that stress (e.g., caused by lack of water) slows down the increase in fibrousness and consequent decline in digestibility which occurs as forages mature (Rhoades 1983). Grasses around permanent water, in the heavy soils of river beds and valley bottoms, have relatively good supplies of soil water. In the Amboseli region, and perhaps elsewhere, these grasses are of lower quality than the grasses of the wet-season dispersal areas (average crude protein 5.0% and 6.2 respectively, Western 1975).

In the Camargue study, for most of each year, no part of the horses' range was more than $\frac{1}{2}$ km from water, and they never had more than 2 km to walk. The availability of water therefore did not play an important role in determining the choice of foraging habitat by the horses. Each month, the quality of the herb layer was monitored, as described in Appendix 4, using chemical measures (% crude protein, phosphorus, ADF and energy) and physical ones (the abundance of the best quality components, the green tissues of grasses and forb, g/m^2). The time spent feeding in each habitat varied seasonally; see Fig. 12. The principal habitats in the warm season (April–September) were the Marshes and the Fallow Fields, which have botanical affinities with the Marshes (Bassett 1978). In the cool season (October–March), most of the horses' feeding time was spent on high-

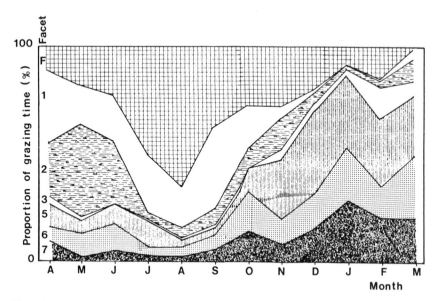

Fig. 12. Seasonal changes in the use the horses made of their habitats for feeding. Facets F, 1, 2, 3 were low-lying, marshy; Facets 5, 6, 7 were on higher ground. Facet 4, Sansouïre, was never used for more than 1% of the time. (From Duncan 1983, courtesy of Blackwell Scientific Publications.)

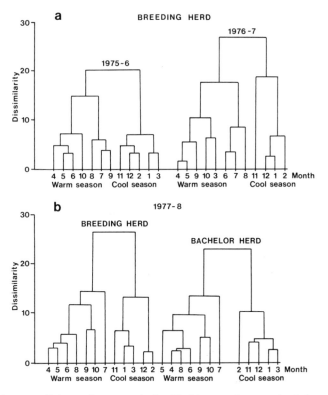

Fig. 13. Average linkage cluster analysis of habitat use by month of the year.

er ground in Enganes and Grasslands. This seasonal pattern was repeated each year and in each herd (see Fig. 13), though March and October grouped with the warm seasons when growth began early or ended late. Within the major seasons, the months did not group consistently.

The horses ranged widely, using all the land facets to some extent each month; yet they were highly selective, particularly in summer, when the best food items were most abundant (Duncan 1983). The use made of a facet was unrelated to the facet's area in the warm season (see Table 6), and even in winter, the areas of the facets accounted for only a half of the variance in the use the horses made of them.

As an index of habitat selection, the use (% feeding time spent) made of a facet was corrected by the area it covered (Hunter 1962) and log-transformed to normalize it. The correlations between this preference index, P_2,[d] and a wide range of parameters of the physical and chemical composition of the herb layers of the facets are shown in Table 7.

---

[d] $P_2 i = \log_{10} (\% \text{ time spent} / \% \text{ area covered} + 1)$ for a given facet (i)

Table 6. The Proportion of Variance in the Use of a Habitat for Feeding (U_i, % of total feeding time) that is Accounted for by the Area of That Habitat

Month of 1975–1976	April	May	June	July	August	September	October	November	December	January	February	March
Variance in U_i (%)	2	7	5	2	8	12	0	23	64*	50*	51*	50*

*$p < 0.05$.

Table 7. Correlation Coefficients of the Transformed Index of Preference (P_2), with Parameters of Sward Composition and Biting Insect Numbers

	Chemical			
Month	Protein	Phosphorus	Fiber	Energy
4	0.473	0.542	0.051	0.209
5	0.228	0.176	−0.050	0.006
6	0.324	0.363	0.247	0.238
7	0.725*	0.808*	0.576	0.549
8	0.615	0.773*	0.433	0.428
9	0.482	0.478	0.618	0.462
10	0.634	0.680	0.700	0.695
11	0.179	0.167	0.311	0.358
12	0.495	0.660	0.660	0.581
1	0.336	0.382	0.557	0.467
2	−0.028	0.108	0.274	0.137
3	−0.139	−0.263	0.074	−0.012

	Physical					
Month	Green grass	Green dicots	Green phytomass	Green (%)	Total grass	Grass (%)
4	0.871**	−0.060	0.779*	0.755*	0.691	0.964**
5	0.776*	0.183	0.742*	0.495	0.811*	0.780*
6	0.735*	0.250	0.783*	0.716*	0.741*	0.686
7	0.645	0.751*	0.795*	0.790*	0.495	0.526
8	0.457	0.835**	0.767*	0.688	0.515	0.406
9	0.793*	0.654	0.909**	0.921**	0.377	0.508
10	0.879**	0.660	0.842**	0.614	0.277	0.550
11	0.800*	0.793*	0.800*	0.123	0.682	0.495
12	0.906**	0.840*	0.879**	−0.155	0.380	0.167
1	0.769*	0.444	0.616	0.793*	0.284	0.483
2	0.908**	0.536	0.533	0.775*	0.184	0.202
3	0.366	−0.394	−0.241	0.053	0.658	0.789*

	Physical			Insects		
Month	Halimione	Halimione (%)	Total phytomass	Mosquitoes	Tabanids	Midges
4	−0.351	−0.353	0.171	0.521	i.n.	i.n.
5	−0.321	−0.330	0.402	0.854**	−0.112	−0.007
6	−0.339	−0.343	0.632	i.n.	0.173	0.084
7	−0.228	−0.268	0.400	0.645	0.896**	i.n.
8	−0.293	−0.322	0.562	0.843**	0.227	i.n.
9	−0.352	−0.372	0.229	0.700	0.559	i.n.
10	−0.311	−0.284	0.193	0.780*	i.n.	i.n.
11	−0.129	−0.157	0.748*	i.n.	i.n.	i.n.
12	0.124	0.130	0.514	i.n.	i.n.	i.n.
1	0.126	0.134	0.329	i.n.	i.n.	i.n.
2	0.022	0.037	0.476	i.n.	i.n.	i.n.
3	0.076	0.160	0.384	i.n.	i.n.	i.n.

Source: Data from Duncan 1983, courtesy of Blackwell Scientific Publications.
*$p < 0.05$; **$p < 0.01$; i.n., insufficient numbers.

Areas with high densities of biting-flies were not avoided, though they were used more intensively by night than by day (see Duncan 1983, Fig. 2). Of the chemical parameters, there were significant positive correlations with the average values of crude protein and phosphorous in the swards. The physical parameters were, however, much more predictive of the preferences of the horses: in particular, green *phytomass* (the biomass density of grasses and forbs) was significantly and positively related to the index of preference in every month from April to December.

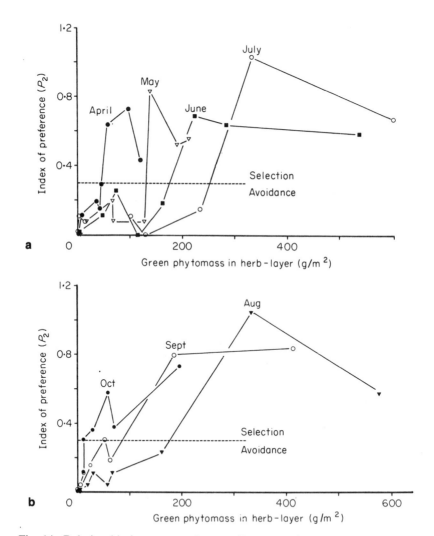

Fig. 14. Relationship between preference (P_2, see text) for a facet and the green phytomass in its herb layer. (From Duncan 1983, courtesy of Blackwell Scientific Publications.)

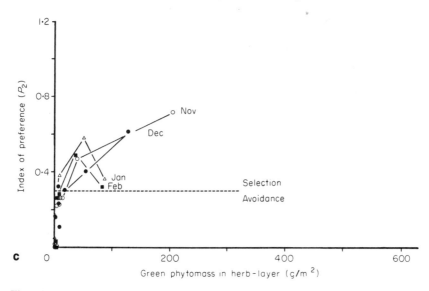

Fig. 14. *Continued*

Closer examination of this relationship showed that it was nonlinear, and tended toward an "S" shape; see Fig. 14 a, b, and c. As green phyto-mass increased from zero, there was initially little change in P_2, then a rapid rise. At high phytomasses, there was little further change in P_2.

The horses therefore avoided facets with green phytomass densities below a specific threshold in each month; facets with densities above the threshold were preferred. Defining the threshold as the value of green phytomass for which $P_2 = 0.3$ (i.e., log 1.0 is no avoidance or selection), the regression of the threshold value on the maximum value of phytomass in any facet was calculated. The threshold was closely and positively related to the maximum amount of green phytomass available in any facet ($r = 0.94$, $b = 0.393$, $n = 11$, $p < -0.001$): as the horses' best food items became more abundant, so they required a higher density of green forage before they would prefer a facet.

This result confirms the prediction that the horses would prefer areas with high densities of green plant tissue. Interestingly, the threshold at which a facet became preferred changed in a manner consistent with optimal foraging theory—as the abundance of food increased, so did the threshold (Pyke et al. 1977, Krebs 1978).

During the remainder of the year (January–March), correlations between preference and herb-layer characteristics were weak, and in view of the small amount of data (3 months × 8 facets) and the fact that there were only minor changes in habitat use, the results for the 3 months were pooled. There was no significant correlation of preference with green phytomass or green grass, but a highly significant one on total herb layer (see

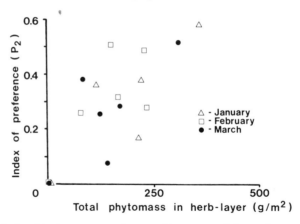

Fig. 15. Relationship between preference (P_2, see text) for a facet and the total phytomass in its herb layer during the latter part of the cool season: $r = 0.71$; $p < 0.01$. (From Duncan 1983, courtesy of Blackwell Scientific publication.)

Fig. 15) for the facets in which some green matter was present (3–7, F). The aerial parts of wetland geophytes (e.g., *Phragmites*) die off in winter, and the horses used the wetland habitats (1, 2) only rarely and dead reeds never (see Chapter 4).

When green plant matter was scarce (in winter, when there was less than 90 g/m² in any facet), the horses changed strategy and broadened their preferences to include the less nutritive dead plant parts of non-geophyte perennials. Instead of searching out areas with the maximum amount of high-quality food (green), they searched out the areas with the greatest total amounts of non-geophyte herbaceous plants. Presumably at 90 g/m², the rate of intake of green matter was so low that the horses could not maintain an adequate daily intake, and it was better to eat more of a poorer diet. Again, this behavior conformed to a prediction of optimal foraging theory and underlines the importance for horses of maintaining a high-intake rate.

Comparisons with Grazing Bovids

Data on Camargue cattle are available from near this study area. These show that cattle have a very similar pattern of habitat use but show only a weak preference for marshes, unlike the horses. The preferred habitat of the cattle was the grasslands in fallow fields in all seasons except the winter, when the herb layer was completely eaten out (Ferrazzini 1987).

In a Canadian boreal forest, horses were more ubiquitous in their use of habitats than were sympatric ruminants. Their behavior was more similar to cattle than to elk or deer (Salter and Hudson 1980).

In a detailed study of the use of habitat by cattle and horses in the New Forest overlap was 0.58 to 0.89 in the four seasons of the year (Putman 1986, Putman et al. 1987). Both species were extremely selective, spending respectively 34% and 54% of their feeding time on the grasslands, which covered only 7% of the area. The horses showed a flexible seasonal pattern of habitat use, switching among habitats. The cattle, however, showed little seasonal variation, perhaps because of the habit of providing supplementary fodder in winter on the main grasslands. In this study the horses used the closely cropped, high quality grasslands to a greater extent than cattle. This result seems to be a general one in temperate ecosystems (Gordon 1989a,b,c): on a Scottish moorland the feeding strategy of cattle led them to move onto the coarse, high biomass grasslands at times when the horses were still using short, higher quality swards.

Studies of tropical grazing systems have shown that zebras, in common with grazing bovids, have a pattern of catenary movement which is exactly the opposite of the pattern described here for Camargue horses. At the height of the growing season, all the tropical grazers move to the edaphic short grasslands at the top of the catena (Bell 1970, Maddock 1979), while the Camargue horses move down the catena to the tallest herbaceous plants in the marshy sumps. The movements in each grazing system are reversed at the end of the growing season.

The different pattern of movements underlies the flexibility of the feeding behavior of ungulates. The feeding strategy of the zebras did not necessarily differ from that of the horses. If the dry season strategy of zebras is to seek out the highest plant biomasses, then this would naturally take them to the catena bottoms in this season.

The movements of zebra resemble closely those of the most similarly sized bovid, the wildebeest (Bell 1970), but zebra tend to precede the wildebeest into the catena sumps, where the grass is longer, which implies that zebra require taller grass than do wildebeest to maintain their intake rate. This is consistent with data from Amboseli (see Fig. 11), which show that zebra move further away from water than do wildebeest, presumably to areas with taller grass. Why the movements of equids relative to bovids differ in tropical and temperate ecosystems is unknown.

On the Predictability of Habitat Use

A strongly selective strategy is nutritionally optimal when the differences in resource quality among habitats are predictable. Where resource quality changes rapidly and unpredictably, less selective or unselective habitat use is indicated. Plant communities and their location and nutritional value are highly predictable, and the horses were selective in each month. For most of the year, no significant part of the variance in their use of habitats was accounted for by the area of the habitat. They ranged widely, perhaps in order to monitor the state of the resources, but their strategy was strongly selective. As predicted, the horses preferred the habitats where the best food items were most abundant.

The nutritional bias of the horses' strategy of habitat selection is underlined by the fact that (a) they became more selective as the abundance of food resources increased, and (b) they widened the range of preferred items when resources were rare, in winter.

Biting flies affected the use of habitat strongly for nonfeeding (see also Keiper and Berger 1982), but not for feeding activities. Feeding horses and tabanids (Bailey 1948) are apparently both attracted by high densities of green, growing plants.

There are suggestions that wild equids may compromise nutritionally optimal strategies to a greater extent. At night, when the risk of predation is high, Plains zebras have been reported to move to short grasslands, both in Ngorongoro Crater in Tanzania, and in the Amboseli region (Klingel 1967, Gakahu 1982).

In the Mountain Zebra National Park, where groups of females show very strong habitat preferences, the preferences of bachelor male zebra were weak and statistically nonsignificant (Penzhorn 1982b). Different foraging strategies are well known in male and female Cervidae (see Clutton-Brock et al. 1982, page 247), where the (dimorphic) males and females have different food requirements. Equids are monomorphic and the two sexes do not differ in their food preferences (see Chapter 4). The different strategies of bachelor and breeding zebras imply that nutritional considerations were paramount for breeding groups, with lactating females which have high nutrient requirements. Bachelors with lower nutrient requirements have different patterns of movement, which are likely to be suboptimal nutritionally, but optimal in terms of seeking mates. In the Camargue herd from 1979, the bachelors followed the breeding herd around, but in the breeding season, they spent less time feeding than did the females. This compromise with their reproductive strategy had consequences for their nutrition.

Conclusions

The three predictions (pages ■–■) are, in the main, upheld by what is currently known of the foraging behavior of equids:

1. Daily foraging time, 14–15 hr, is close to twice as long as the median values in cattle and sheep (see Fig. 5). Variations in foraging time rarely exceed ±10% of this value; in cattle and sheep, grazing times are much more variable.

2. The temporal organization of foraging contributed to a high rate of nutrient assimilation by

   a. Avoiding long periods without feeding

   b. Feeding to a satiety threshold, probably to maximum gut fill.

3. Selection of feeding habitat by Camargue horses was consistent with maximizing their nutrient assimilation rate:

   a. They preferred, for most of the year, the habitats in which the best-quality food (green tissues) was most abundant (Fig. 14)

   b. When the amounts of green phytomass increased, the horses became more selective, and vice versa

   c. When green phytomass became very sparse, in winter, they broadened the range of preferred items to include dead tissues, apparently in an attempt to maintain their rate of intake (Fig. 15).

However, for each prediction, there are circumstances when the nutritionally optimal strategy is compromised to some extent:

1. Foraging time is never maximal and is always constrained by resting, which occupies about a quarter of the 24 hr. Reproductive behavior in adult males appears to compete with feeding and to cause an 8% decline in foraging time during the breeding season. This compromise led to a decline in the quality of the diet of the males.

2. When biting insects are abundant, the animals alter the temporal patterns of foraging behavior: homeostatic mechanisms regulating gut fill break down, and even the fine structure of feeding behavior is disrupted. However, total feeding time is only slightly reduced.

3. Antipredator behavior may cause the use of suboptimal feeding habitats by day, when tabanids are active in the Camargue; however, by night, the horses concentrate their feeding in the preferred habitats to such an extent that habitat use on a 24-hr basis may be unaffected.

Section C   Nutrient Use

Chapter 6   Social Organization, Mating System, and Feeding Behavior

> With white tails smoking free,
> Long streaming manes, and arching necks, they show
> Their kinship to their sisters of the sea
>
> Roy Campbell
> from "Horses on the Camargue"

Horses are highly social ungulates, yet life in groups has inherent costs: it increases both the competition for food and the transmission of diseases (Alexander 1974), to cite only two. In other social species, grouping can allow animals to exploit their food resources better (e.g., large carnivores, Schaller 1972, Kruuk 1972; frugivorous primates, Wrangham 1980; birds, Wittenberger and Hunt 1985) and to reduce their chances of being predated (Treisman 1975, Pulliam and Caraco 1984, Mangel 1990).

Ungulates, especially grazers, do not usually group to defend their food resources, but they commonly do so to avoid predators (Bertram 1978). Nonetheless, the pattern of distribution of the food appears to impose limits on the size of their groups (Jarman 1974). When the food resources occur in sparsely distributed small packets, such as browse buds, the small antelopes which eat them live singly or in small groups; however, grazers such as wildebeest, which feed on large patches of homogeneous grasslands, often occur in herds of hundreds or thousands.

There has so far been no quantitative analysis of the effect of food abundance on group size in equids as there has been for some other species (e.g., canids, Moehlman 1986, bighorn sheep, Berger 1988). Nonetheless, there are some broad patterns of association between equid group sizes and habitats, which suggest that the abundance of their food supplies has a strong effect on the size of the groups they live in. In near-natural ecosystems with abundant food and predators, zebras group like wildebeest in temporary assemblages, which may number thousands of individuals (Klingel 1967, Maddock 1979). Where food resources are sparse, in heavily grazed ecosystems or deserts, equids live singly or in small groups (Tyler 1972; Moehlman 1974; Woodward 1979). These broad patterns of grouping are probably determined, as in other species, by the females (e.g., primates, Wrangham 1980), whose reproductive success is more strongly affected by their ability to acquire nutrients than it is in males (Gosling 1986, Clutton-Brock 1989).

Males, whose reproductive success is determined principally by their opportunities to mate females, have two broad options: (1) to sit and wait, defending part of the females' ranges against other males, or (2) to tend individual females, and thus acquire priority of access to these (Gosling 1986, Clutton-Brock 1989, Rowell 1988). Male equids adopt both of these mating systems. Early studies showed that in the desert-living species (Grevy's zebra and two species of asses), the dominant males are territorial,[a] while in horses and in Plains and Mountain zebras, which typically live in richer mesic habitats, the males tend small harems of females (Klingel 1975). The fact that each mating system is associated with a particular type of habitat suggests that, as in other ungulates, the males adopt different mating systems according to the patterns of distribution of the females.

Klingel (1975) argued that equids have only these two mating systems, that each species adopts only one, and that the territorial system is maladaptive in the desert environment. The third opinion was based on the view that the animals would do better to adopt a harem system in a biome where the distribution of resources is unpredictable.

In a synthesis of more recent studies, Rubenstein (1986) disagreed:

> Horses and zebras exhibit few niche or phenotypic differences yet they display a wide range of social systems. The social diversity seen on Shackleford Banks is mirrored in other feral equid populations. Female horses associating with territorial males are found in Exmoor, on Cumberland Island and in the New Forest, and female asses living in closed membership harem groups occur on Ossabaw Island.

The data for the asses are interesting but need confirmation, for they come from one small and disturbed population of 75–80 individuals, in which all the males were vasectomized at the start of the study.

In an elegant study of the other desert equid, Grevy's zebra, Ginsberg (1988) has shown that males fine-tune their behavior toward individual females in response to their patterns of movement and to the probability

---

[a] Note that they do not exclude other males, but simply dominate them on territory.

that they will stay on the male's territory. Some of these females spend 3 months with the same male, but they have not been found to form the long-lasting female-male ($♀-♂$) bonds of the mesic-habitat equids.

Among the mesic-habitat species, where males typically tend females, territorial males have been found only in special circumstances, where the costs of territorial defense are exceptionally low, on barrier islands where there are few adult stallions, and in managed domestic herds with sex ratios strongly biased in favor of females (Rubenstein 1981, 1986).

In a review of studies of feral horses and of Plains and Mountain zebras, Berger (1988) has found that of 21 studies of mesic habitat equids, the mating system is based, in all cases but the one, mentioned above, on long-term harems with members who have strong social bonds, both $♀-♂$ and $♀-♀$.

The relative inflexibility of the social organization of harem-forming zebras is clearly demonstrated by two recent studies. Berger (1988) manipulated the distribution of the food supply of Grevy's and Mountain zebras in the same enclosure. The Grevy's males always set up territories, while the Mountain zebras always tended females. The female Grevy's zebras modified their grouping behavior as the food supplies changed, but the Mountain zebras did not.

The second study was conducted in the Samburu area of Kenya. Grevy's zebra are common there, and they have a territorial mating system. The few Plains zebra, which are on the edge of their range in this semiarid region, lived in the characteristic stable harems of this species until they experienced a drought so severe that adult mortality approached 25% and juvenile mortality 70%. The harems, not surprisingly, broke up, but even then, the males never switched to a territorial system (Ginsberg 1988, page 204). The social organization of this species, within the range of habitats and demographic structures encountered in wild populations is therefore less flexible than Grevy's. These studies provide evidence that Plains zebras stallions always tend females on a long term basis.

In conclusion, in habitats where food supplies are sparse, female equids live in small groups or singly; when food supplies are abundant, they may live in large groups, which include many harems. Recent studies confirm that males of the desert species are flexible in their social organization and mating systems. Males of the mesic-habitat species, on the other hand, always tend females on a long-term basis. They defend territory only in circumstances so exceptional as to be of marginal relevance to the biology of their populations.

These conclusions led us to expect in the Camargue study that

1. The males would establish harems and maintain long-term bonds with individual mares
2. As the population increased relative to the food resources, feeding competition would lead the females to live in small groups.

In this chapter, I describe the social behavior of these horses during the process of feralization, and I then test these hypotheses. Harem-based social systems have been little studied: In particular, there is little direct information on the extent to which the social system corresponds with the mating system. The accessibility of this herd allowed this information to be obtained: methodological details can be found in Appendix 7.

Social Organization

Group Sizes and Home Ranges

From the beginning (December 26, 1973), the horses, initially 14 (1 ♂ and 6 ♀ adults aged 4–18 years, 2 ♂ and 1 ♀ 2-year-olds, 2 ♀ and 2 ♂ yearlings) split into two groups,[b] a breeding group and a solitary 2-year-old male. The bachelor unit was formed in 1974–1976 by nine young males, mostly chased by the oldest stallion (Darius, born in 1969) from the breeding unit. Five females joined the bachelors (1976–1979), usually when they were in estrus.

During 1977, the three oldest bachelors, ages 4–5 years, returned to the original breeding group, and after considerable fighting over a week, they divided this group into two harems and established an independent group. The following year, another (4-year old) stallion set up an additional group. At various stages in the study, three groups therefore lived separately from the original breeding group (D group): the Bachelors, the G4 group and I4 group.

In the first year (1974), the D group used the whole pasture (Fig. 1, a, D group), and the solitary 2-year-old bachelor appeared to pose no threat to the herd stallion. After the expulsion in 1975 of a 3-year-old (G4) who had mated with many of the mares in that year, the home range of the D group was restricted to the north and east until 1976; see Fig. 1, b–c. The part of the range most intensively used was the Fallow fields. The home range of the Bachelors was the south and southwest in the first 3 years. The use of space during this period appeared territorial because when the bachelors approached the home range of Darius's group, he chased them away, and the bachelors were often seriously bitten.

The pattern changed markedly in the subsequent years (see Fig. 1, d–e), when the ranges became overlapping. The other two groups, which have at some time been independent of the D group (G4, I4), showed much the

---

[b]The horses were considered to be in more than one group when the distance between the two (or more) groups was greater than the diameter of any of them.

Fig. 1. Home ranges of D Group and bachelors, 1974–1978 (a–e).

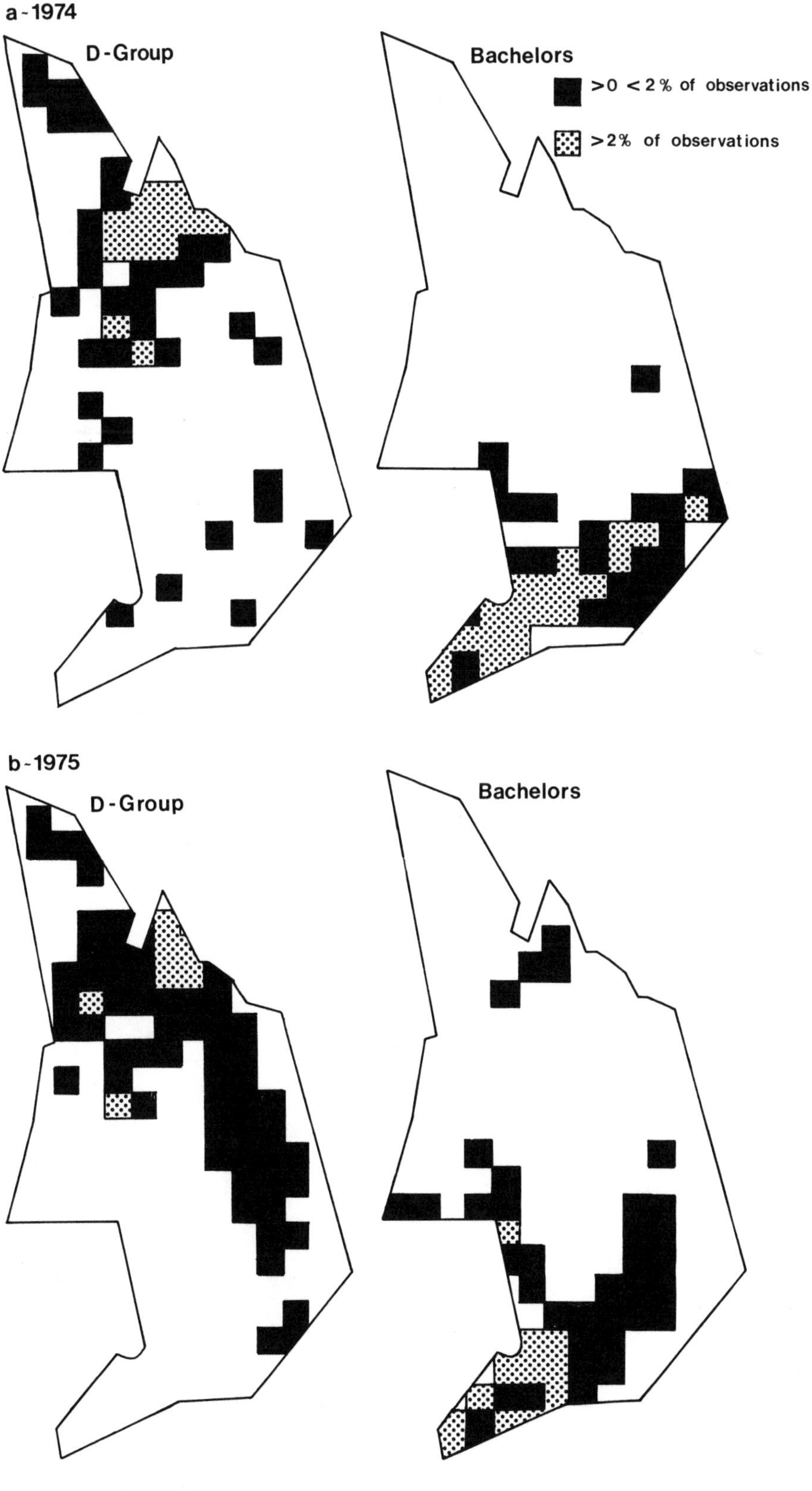

a - 1974
D-Group
Bachelors
■ >0 <2% of observations
▨ >2% of observations

b - 1975
D-Group
Bachelors

c - 1976

D - Group   Bachelors

d - 1977

D - Group   Bachelors

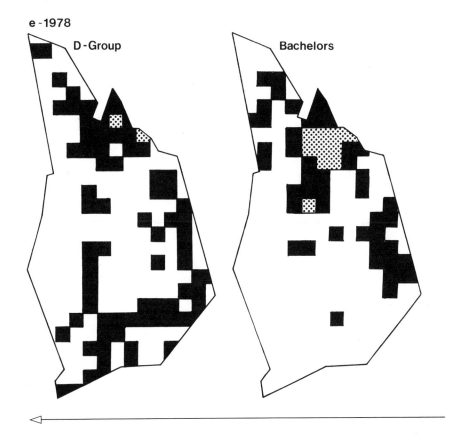

e -1978

D-Group

Bachelors

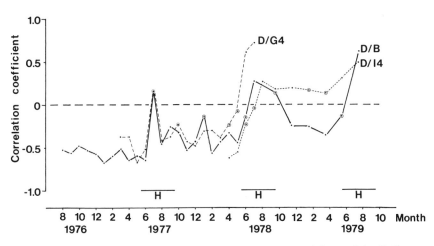

Fig. 2. Correlation between home ranges of new groups and those of the D Group.
H = periods of horsefly abundance; D/B = D group/bachelors; D/G4 = D group/
G4 group; D/I4 = D group/I4 group; circled points are not statistically significant.

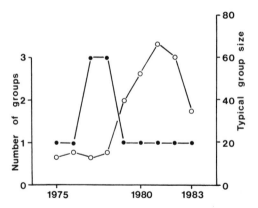

Fig. 3. Variations during the course of the study of the number of groups (filled circles, median) and the typical group size (i.e., the number of individuals ≥ 18 months old; open circles; cf. Jarman 1974).

same kind of behavior as the Bachelors—an initial avoidance of the D group home range was translated into a preference for it. This is shown clearly by the correlation coefficient between the numbers of observations of each herd in each grid square for August 1976–December 1979; see Fig. 2. The initially negative correlation became briefly nonsignificant in the summer of 1977 for both groups. In summer 1978, the three groups overlapped extensively with the D group, with which G4 group then associated permanently. The other two remained independent until the following summer: from December 1979–September 1983, all the horses remained together 90% of the time, in a single large group.

From the information above it is clear that there was some tendency for the family units to split into small groups until 1979, and group size remained low; see Fig. 3. However, as the food supply became very sparse and the horses nutritionally stressed (1981–1983, see Chapter 7, Fig. 2), all the horses associated in one large herd of >30, which contained seven or more harems with their stallions. The size of this group declined in the last 2 years of the study only because there were fewer horses in the population (see Chapter 1, Fig. 6).

Spatial Relations within the Large Herd

The basic unit in horse societies is composed of a mare and her young offspring: this matrilineal unit maintains close spatial relations in all activities (Tyler 1972, von Goldschmidt-Rothschild and Tschanz 1978, Wells and von Goldschmidt-Rothschild 1979) and lasts until the second or third year, when the young horses become reproductively active or join a bachelor group. When not with these matrilineal units, young horses spend most of their time with their peers. There was no suggestion that at this level

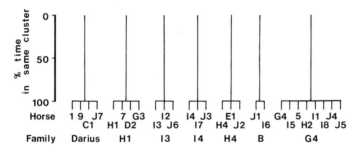

Fig. 4. Single linkage cluster analysis dendrogram linking the horses that are 4+ years of age, according to the percentage of periods in which they were in the same cluster. An SLCA was done for each period of the year separately (January–April; May–June; July–August; September–October; and November–December and the horses were scored as belonging to one of seven clusters. (From Duncan et al. 1984a, courtesy of Academic Press.)

horses tended to associate with close relatives (Wells and von Gold-schmidt-Rothschild 1979).

By 1979, the horses were bonded into a social system characterized by stable spatial relations between stallions and harems of adult mares. There were seven such family units; see Fig. 4.

The composition of the families by age and sex on December 31, 1979 is shown in Table 1. In each, there were one or two (G4, Bachelors, which have now become a breeding family) socially mature males which herded the females. Six of the families had 5–11 members, and one (G4) was much larger, with 28 horses. The adult members have remained in these groups to the present (1990), unless they died or were removed. These family units moved about together as members of the large group in much the same way as a hamadryas baboon clan. In spite of the spatial proximity of so

Table 1. Composition of the Family Units on December 31, 1979, by Age and by Sex

Name	≥5 yr		4–5 yr		3–4 yr		2–3 yr		1–2 yr		<1 yr	
	♂	♀	♂	♀	♂	♀	♂	♀	♂	♀	♂	♀
Darius	1	2[a]	1		3		2			1		1
G4	2	4[a]		2[a]	1	3[a]	2		3	2[a]	6	3
H1	1	3[a]								1	2	1
H4	1	1[a]		1[a]				1[a]	1			2
I3	1	1[a]		1[a]			1		2		1	
I4	1			2[a]				1[a]	1	1[a]	1	1
Bachelors	1		1			1[a]		1[a]		2[a]		

[a] Breeding females.

many males, the probability of paternity for holders of one-male harems was high (95%, $n = 20$, years 1977–1978; data in Fig. 4, Appendices 7, 8). The mating system therefore corresponded closely with the social system.

Social Dynamics

In harem-based equid societies, social relations of adults are very stable (Berger 1986), so the dynamics concern primarily the young animals.

In the early years (1974–1976), some of the young males left their matrilineal units aged 2–3 years and they formed consortships with and mated with young and even adult females within the D group. The eight oldest ($2\frac{1}{2}$–$3\frac{1}{2}$) left after being severely chased and bitten by the stallion, who was not their father. The ninth, the only son of the stallion, left without being attacked, aged $1\frac{1}{2}$ years. After the development of a complex social structure within the large group, some of the young males left their natal families, mostly with no visible aggression from their fathers, to attach themselves either to bachelor groups or to young females (C. Feh and A.M. Monard personal communication, 1988).

Young females also transferred out of their natal families. Some of these females transferred permanently to other families before foaling, usually when about 2 years old, while others transferred after the birth of their first foal; see Table 2. It was obvious that mares were much easier for stallions to abduct when hampered by a slow-moving newborn foal. The stallion of the natal family made little or no attempt to keep strange males away from such young females, though they did do so when any of their adult females foaled.

Once the original herd had split into a number of harems, the social dynamics were therefore closely similar to those of wild herds: virtually all the young animals emigrated from their natal bands between ages $1\frac{1}{2}$ and $3\frac{1}{2}$ years (c.f. Penzhorn 1984, Kaseda et al. 1984, Berger 1986). In many species a prime function of dispersal of young individuals seems to be inbreeding avoidance (Packer 1985): the particular circumstances of this herd make it possible to show that the avoidance of incestuous matings was not simply a product of the social dynamics.

Inbreeding Avoidance

The average inbreeding coefficient of the horses remained below 10% and showed no tendency to increase in spite of the lack of genetic exchange with other herds (Duncan et al. 1984a). This occurred because there were few matings between close relatives.

There was clearly an inhibition against mother–son matings: no stallion ever had its mother in its harem. Only two mother–son matings were observed in 6 years (<1% of observations), and none gave rise to a foal. This was in spite of all but one of the young males both staying in their

Table 2. Data on Sexual and Social Affiliations of Young Females Born 1974–1977

| | First foal sired by | | | Transfer out of natal band | |
| | Dominant stallion of natal family | Stallion of other family | Subordinate stallion of natal family | Before first foal | With first foal |
Female identity					
I1			•		at 3 yrs
I2			•		at $3\frac{1}{2}$ yrs
I7		•			at $3\frac{1}{2}$ yrs
I8		•			at 3 yrs
J3		•			at 3 yrs
		(own father)			
J4		•		at 2 yrs	
J5		•		at $1\frac{1}{2}$ yrs	
J6		•			at $2\frac{1}{2}$ yrs
K2		•		at $1\frac{1}{2}$ yrs	
K4	(not known)			at 2 yrs	
K5		•		at 2 yrs	
		(own father)			
K7		•		at 10 mths with mother	
L1		•			at $2\frac{1}{2}$ yrs
L3		•		at 8 mths with mother	
L6		•		at 2 yrs	

Note: None bore foals of the dominant stallion of the family into which they were born; all transferred before they had produced a second foal.

family unit until they were old enough to breed (at age 2 years) and (until 1976) mating with other mares ($n = 7$).

Inbreeding between fathers and daughters was generally avoided because the daughters transferred out of their natal families. However, in this herd, where in the early stages of the study, young males successfully mated mares, were expelled, and then reappeared as harem stallions, this mechanism of incest avoidance was not foolproof, and mares J3 and K5 bore foals of their father (G4; see Appendix 8). There is therefore no inhibition against father–daughter matings per se. When young mares bred before transferring, they were mated by subordinate males of the D group in the early years, and in later years by males of other groups; see Table 2. Sexual activity between them and the dominant stallion of their natal family at the time of their birth, regardless of whether he was their father was uncommon; see Table 3. This phenomenon is being studied further by A.M. Monard.

These observations make it clear that horses cannot assess their degree of relatedness, as in some other species (Fletcher and Mitchener 1987, but

Table 3. Matings Involving Young Mares Born into Darius's Group

Mares	Number of matings observed	
	With Darius	With other males
I1	3	77
I2	4	15
I7d	1	14
I8d	1	34
J3	1	17
J4d	0	10
J5d	0	4
J6d	0	14
K2d	0	16
K5	0	3
K7	0	12
L1	0	8
L3d	0	6
	10	230

Note: Data gathered in 1974–1979; there were seven other males; d = a daughter of Darius; data from Duncan et al. 1984a.

see Grafen 1990). Father–daughter matings, where the relationship could not be "deduced" from membership of the same family, were not avoided: young horses appear to be sexually imprinted on the members of their natal family. Similar conclusions were drawn from a long-term study of mustangs in the United States (Berger and Cunningham 1987).

Social Ranks

As in all horse groups, agonistic behavior was frequent, and adult horses always ranked above young ones (Houpt et al. 1978, Houpt and Keiper 1982, Wells and von Goldschmidt-Rothschild 1979). The ranks of mares when adult (age 6 years) were principally determined by seniority: 85% of 36 pairs involving 16 mares were correctly predicted from the birth order. Adult weight predicted rank no better than chance, 20/36.

Dominance relations between adult males and females were more complex. Each stallion herded all of his mares and frequently reinforced this herding with bites, generally to move the females away from competitors which the stallion was not able to chase far or fast enough. In this context, therefore, the stallion dominated all the mares. In competition for resources such as food, shelter, and water, young males ranked initially according to their ages (Wells and von Goldschmidt-Rothschild 1979). In 1979, harem stallions, like mares, showed clear dominance over mares younger than themselves (see Table 4, I4/I7, J3, I3/J6, H4/J2, 4–6 threats in each case), whereas with mares older than themselves, dominance rela-

Table 4. Matrices of Head Threats between the adult (> 3 year old) Horses of the Single Male Harems

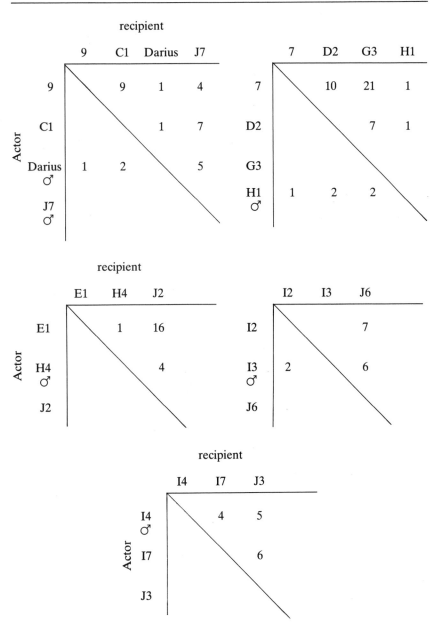

recipient

	9	C1	Darius	J7
9		9	1	4
C1			1	7
Darius ♂	1	2		5
J7 ♂				

	7	D2	G3	H1
7		10	21	1
D2			7	1
G3				
H1 ♂	1	2	2	

recipient

	E1	H4	J2
E1		1	16
H4 ♂			4
J2			

	I2	I3	J6
I2			7
I3 ♂	2		6
J6			

recipient

	I4	I7	J3
I4 ♂		4	5
I7			6
J3			

Note:Data gathered during 1979, $58\frac{1}{2}$ hours of observation; actors are ranked by age; the letter indicates the year of birth (1974 = I).

tions were not clear: there were fewer threats (0–3) and some reversals (H4/E1; H1/7, D2; Darius/9, C1; I3/I2). In all of these cases the male had been dominated by the older females prior to his emergence as a harem stallion (Wells and von Goldschmidt-Rothschild 1979 Table 3).

The adult stallions often interacted in ritualized displays involving nose–body contacts, smelling of the flanks and genitals, and defecating, usually on existing dung piles (Tschanz 1979, Turner et al. 1981). The function of these displays is not known, but it is perhaps a means by which they monitor each other's physical condition. During the displays, the stallions generally keep between their mares and the other stallion(s). In displays involving adult stallions of different families, the order in which the individuals leave is variable and depends on such factors as the distance their females have moved since the display began (Feh, 1987). However, in displays between males of the same family, adult and subadult or dominant and tag males, it is almost invariably the dominant[c] horse which defecates last (Tschanz, 1979; Feh, 1987).

The adult stallions in different harems head-threatened each other so rarely (0.04 threats/horse/hr) that it was not possible to demonstrate the existence of an interfamily hierarchy in the field. There was no limited, defensible resource as in the drinking site in the Grand Canyon (Berger 1977). However, in 1989, the herd was experimentally provided with limited water supplies, and a linear hierarchy among adult males was found (C. Feh, personal communication, June 1990).

On the Feralization of the Social Organization

As predicted, the stallions of this herd tended the same mares on a long-term basis. For 2 years, there was a pattern of use of space, which looked territorial. As in other cases where equids of mesic habitats have been shown to defend territory, the dominant male tended a harem. Further, the costs of defense were unnaturally low because the breeding sex ratio ($\male\male > 4$ years, $\female\female > 2$ years) was initially only 13% male. When it became 40% male, the territorial defense disappeared.

Within the families, social structure was similar to that described for other feral horse societies. The horses which associated most closely were matrilineal units: the breeding mares with their nonreproductive offspring, foals, yearlings, and 2-year-olds (Wells and von Goldschmidt-R., 1979). The persistence of these bonds beyond weaning implies that the young horses obtained some benefit from continued association with their mother. There is no direct evidence for such benefits, but they probably include access to shelter and food resources. Recent work has shown an

---

[c] Assessed by success fights and by the number of mares mated.

effect of mother's rank on the reproductive success of males, which is un-related to their body size (Feh 1990).

Stable dominance hierarchies closely linked to seniority, not body weight, were found in all groups, at all stages, as in many other horse societies. Rank in artificially created groups of domestic horses has been found to be loosely correlated with body weight (Houpt et al. 1978), but "the determinants of social dominance in horses remain obscure" (Houpt and Keiper 1982, p. 949). In this, natural herd seniority (i.e., age) was the main determinant.

In two key studies of feral herds, it has been stated that ranks are rarely expressed, unstable (Feist & McCullough, 1976), and "of little biological importance" (Berger 1986, p. 159). It has been suggested that the domi-nance of males over all females in feral horse societies somehow suppresses the establishment of clear hierarchies among females (Feist and McCul-lough 1976). This explanation for the lack of linear hierarchies in their data is not convincing for one general and two specific reasons. There is no reason why dominance hierarchies, for the efficient resolution of conflicts over resources, should be less strongly favored by selection in resource-limited than in domestic populations. The specific reasons are

1. At the start of this study (1975), the young stallions (<4 years) ranked for resources according to their ages (Wells and von Goldschmidt-Rothschild 1979) and were dominated by older mares; herding was seen but rarely. Four years later, these males showed the same kind of domi-nance over their females as is found in feral horses. However, there were still clear female dominance hierarchies, and the relative ranks of the adult females remained unchanged with only one exception. The emergence of harem stallions had no strong effect on the dominance relations of the mares.
2. Plains zebras have a harem-based social system, yet the females have linear dominance hierarchies (Klingel 1967).

The likely explanation is that hierarchies are stable only in stable groups. In the mustang societies, some 10% of females >4 years of age changed groups annually in the Granite Range (Berger 1986); in the Camargue, the animals knew each other from birth, and none changed family between 1978 and 1983. The biological significance of dominance in the Camargue herd is evaluated in Chapter 7.

Contrary to the prediction on page 131, the females did not reduce their group size as the food supply declined to the point where their offspring were growing at far less than their potential rate, and they themselves had become very thin (Chapter 7). This was in spite of the fact that the herd, which numbered nearly a hundred in 1982, had split into harems typical of equids of mesic habitats, which elsewhere live separately (e.g., Berger 1986).

The new groups ceased avoiding the D-group home range during the summers, in each case. Two factors were probably involved in this increased overlap: the fields, in the center of the D-group range, were by far the most intensively used facet in the summer; see Chapter 5, Fig. 12; grazing conditions in the fields may thus have attracted the other groups at this time. The other factor that probably played a role was the biting flies: these peaked in June or July, and the members of the small, isolated groups had many more horseflies attacking them than horses in the main herd (Duncan and Vigne 1979). The small groups started associating with the large one in the summer of 1978.

The winter separation finally broke down when the dominant male of the other groups was socially mature (ages 4 years, J1; 5 years, I4; or 6 years, G4). It seems probable that a period of time was necessary for the establishment of bonds among horses and for the young stallions to develop their capacity to repel competitors.

This interpretation is supported by the results of a recent experiment where four stallions were placed on the same pasture with six unfamiliar mares. Two harems were established within an hour; the stallions initially kept the harems apart in exclusive home ranges, but after 3 weeks, they formed a single herd. During the time they were separate, all stallions spent much more time alert than when they were together (C. Feh personal communication, June, 1990).

It was my impression that when the harem of one of the socially mature stallions strayed away from the herd, it was usually the stallion which herded the mares back, and Claudia Feh's observations that stallions were much more vigilant when alone than when grouped suggests that the stallions benefited from communal defense of their harems against competing stallions from other groups. The same argument has been advanced to account for associations of harems of Plains zebras (Ginsberg 1988). I suggest that in the case of the Camargue herd, the families associated because of the flies in the summer and were kept together by the males in the winter. I now examine the possibility that living in the large group increased the feeding competition among the mares.

Feeding Competition

Most of the agonistic behavior observed occurred when the horses were feeding; see Table 5. Some of these threats were clearly related to the acquisition or defense of food items, though many did not seem to concern a particular resource item but seemed rather to be aimed at maintaining individual distance during the meal.

The number of horses that mares tolerated close to them was related to the abundance of green plant matter; see Fig. 5. When the food supply was sparse, the mares spaced out more to feed. The frequency of threats in-

Table 5. Frequency of Head Threats (Given and Received) per Hour in Meals and Intervals

Mare	9	7	I1
Meal	0.59	0.54	0.30
Interval	0.30	0.21	0.27

Note: Meals occupied 65–75% of the 24 hr.

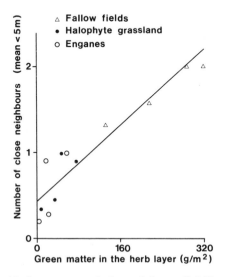

Fig. 5. The relationship between proximity and the availability of green plant matter in the herb layer, $r = 0.94$, $p < 0.001$.

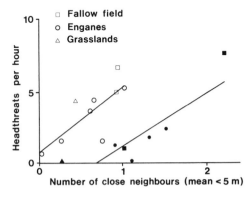

Fig. 6. The relationship between the frequency of head threats and the number of horses in close proximity, for two mares, October–May. Filled symbols = mare 7; open symbols = mare 9. Both regressions are significant ($p < 0.05$).

Table 6. The Mean Numbers of Head Threats per Hour (Given and Received) by the Mares in Small and Large Herds, While Feeding in Two Facets

	Facet	
Herd	Fields	Deep marsh
Small	1.42	0.12
Large	5.0	1.73

creased as the number of horses close to the focal mares increased; the results for the two dominant mares in the cool season is shown in Fig. 6. Because the response of subordinate horses when threatened was to move at least their heads, and usually to walk away, this increase in threat frequency would have the effect of reducing the crowding around the dominant mares.

In 1978, when there were small groups separate from the large one, the foraging of six mares, paired for age and reproductive status, was compared between small and large groups. Mares in the small groups were involved in agonistic interactions very much less frequently (see Table 6), and had more long feeding bouts (see Fig. 7). In the large herd, the majority of bouts were shorter than 1 min.

The horses therefore competed for food, and the mares maintained access to feeding sites and individual distance around themselves by agonistic

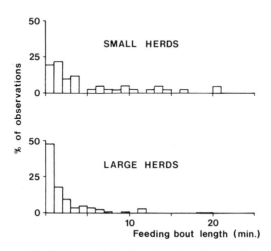

Fig. 7. Frequency distributions of feeding bout lengths for three mares in small herds and three mares of the same ages in the large herd. Mares weighted equally; the distributions are significantly different (k-s test, $p < 0.001$, P. Moncur personal communication, September, 1978).

behavior. Membership of the large herd resulted in a higher frequency of agonistic behavior and the disruption of feeding.

These observations were made at a time of abundant food: as the food supply declined, and feeding competition became critical, even if the mares did not split up into smaller groups, they could have reduced feeding competition by dispersing more widely. This possibility is examined in the next section.

Dispersion of the Horses Within Groups

The average area occupied by weaned horses showed dramatic seasonal changes: during the cool seasons (October to March), the horses occupied some 150–400 m², while in the warm seasons they occupied only 30 m²; see Fig. 8. This was clearly a fly effect, for they sought out other horses, usually close kin, and cooperatively repelled the flies by brushing them off with their tails and bodies. The frequency of such actions by mares rose by a factor of 10 during the fly period (Wells and von Goldschmidt-Rothschild 1979).

During the summers, the horses' dispersion was affected by activity (see Table 7), and there was an interaction between activity and flies: dispersion was affected by flies only when the majority of horses were foraging. Thus, when the flies were abundant, the horses did not disperse for grazing, but remained tightly packed, and a reduction in fly numbers led to an increase in dispersion when most of the herd was grazing.

The patterns of dispersion have also been examined by counting the

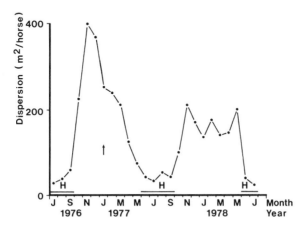

Fig. 8. The dispersion of the herd (average area occupied by weaned horses) in different months of the years 1976–1978. H = horsefly season. The arrow shows the time when intense competition for mares began.

Table 7. The Average Area ($m^2 \pm 95\%$ c.l. where $n > 10$) Occupied by the Weaned Horses in the Principal Habitats at Different Levels of Horsefly Activity

Horsefly index	Facet		
	Grasslands + enganes	Marshes	Fields
a. 20% or fewer foraging			
0–1 (few)	20 ± 11	20 ± 9	25 ± 29
2	20	17	11
3–4 (many)	20	14	20 ± 13
b. >80% foraging			
0–1 (few)	53 ± 20	63 ± 30	48 ± 11
2	42 ± 15	39 ± 10	40 ± 19
3–4 (many)	25 ± 7	30 ± 12	35 ± 14

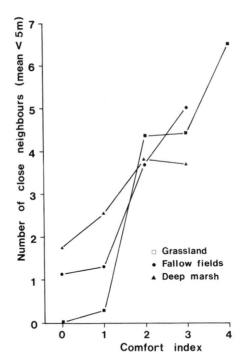

Fig. 9. The relationship between comfort behavior and number of neighbors in close proximity; data from the months June–October only, focal animal foraging.

Table 8. The Average Area (m² ± 95% c.l.) Occupied by the Weaned Horses in the Principal Facets at Different Levels of Foraging Activity, during Cool Seasons

Activity (Percentage of herd foraging)	Facet					
	Enganes		Grasslands		Fields	
	1977	1978	1977	1978	1977	1978
0–20	111 ± 76	61 ± 56	37 ± 58	70 ± 37	82 ± 22	78 ± 45
21–40	336	118	171	96	51	109
41–60	187	145	117	154	71	111
61–80	268	157	313	185	158	120
81–100	311 ± 86	176 ± 24	348 ± 80	172 ± 24	128 ± 16	120 ± 25

Note: 1977 = October 1976–April 1977; 1978 = November 1977–May 1978.

numbers of horses in close proximity (<5 m) to focal individuals. They had many more close neighbors when more biting flies were active, and this effect was evident in all vegetation types; see Fig. 9.

Because the horses spent more time resting and less foraging during the daytime in the warm season, it is possible that the seasonal change in dispersion was simply a consequence of this change in activities during the day (cf. Chapter 5). However, this was not the case, because with 0–20% of the herd grazing, dispersion varied between 37 and 111 m²/horse in the cool seasons (see Table 8) and 11–25 m²/horse in the warm season (see Table 7). The biting flies were therefore responsible for disrupting the normal pattern of dispersion.

To compare dispersion between years, the cool seasons, when flies were absent, are clearly of more interest. The horses were much more dispersed during the early part of winter 1976–1977; see Fig. 8. I interpret the subsequent reduction in dispersion as arising from the intense competition for mares, which began in January 1977: in a competitive situation, the stallions could not allow their mares to disperse too widely. Though less than the fly effect, this social effect reduced the dispersion by about a half (400 m² − 200 m²).

During the cool season, the horses' dispersion was also strongly affected by activity and to some extent by vegetation type; see Table 8. When the majority of the herd was not foraging, dispersion was minimal, whereas when most were foraging, they occupied 120–350 m²/horse. The animals were less dispersed in the fields than in either Grassland or Enganes (two-way ANOVA without replication, $t = 5.2$, $df = 2.274$, $p < 0.01$ for 1977; similar, but nonsignificant trend in 1978), but only when most of the herd was foraging. In 1983, with <80% feeding, the animals' dispersion was less—not more—than in 1978, when their food was abundant, 151 ± 43 m² in Enganes and 168 ± 52 m² in Grasslands.

When there were no biting flies and the food supply was poorest, the horses were tightly grouped to rest and were dispersed to feed, with each weaned animal occupying hundreds of square meters. Feeding mares threatened more frequently when they had more animals close to them, and they tolerated fewer close neighbors if the food was sparser. The increase of dispersion during meals was therefore clearly linked to feeding competition, but the females dispersed less widely after 1976 and showed no increase in dispersion when the food supply was very sparse in 1983. This result was different from what had been predicted, and very different from the kind of flexible patterns of dispersion under variable feeding conditions, which have been found in other species (e.g., sheep, Dudzinski et al. 1978).

General Discussion and Conclusions

When management of this herd of domestic Camargue horses ceased, and males were no longer removed, the herd rapidly developed a social system closely similar to that of wild and feral equids of mesic habitats. The stallions tended mares, and harem membership and dominance relations among adult horses were stable from year to year.

Why do mesic-habitat equids show territoriality so rarely? Territorial defense is time consuming, and it is nutritionally costly to reduce the time spent feeding (Chapter 5). I suggest that the group ranges of mesic-habitat equids are normally too large to be defensible by a male and that the distribution of females in these species is too unpredictable, because water is more abundant than in deserts, for the advantages of territoriality to outweigh the costs. Exceptions to this rule are found only when the costs of territorial defense are lowered (e.g., by management—removal of competitors, or by life in small populations on barrier islands).

The number of breeding females per male in equid harems is remarkably constant over a wide range of habitats and population densities (e.g., 3.1 ♀♀/harem in North America, Berger 1986; 2.0–3.0 in tropical African savanas, Klingel 1975), which implies that the number of females a male can defend is determined by characteristics of the males, not by the optimal feeding group size of the females. Equid males tend on a long-term basis, like many other mammals, because the breeding season is long (March to September in many northern populations, cf. Clutton-Brock 1989). Temporary tending is likely to bring few benefits and to be more costly in time.

In these horses, time for reproductive activities could be taken only out of feeding time (see Chapter 5), and this had immediate dietary consequences. Male horses may be constrained to tend for this reason—an interpretation already advanced for permanent tending in gelada baboons (Dunbar 1978).

An added benefit of permanent tending, for both sexes, is that the male can cooperate with the females to defend the offspring against predators.

Zebra stallions actively defend their families against hyenas (Kruuk 1972) and lions (Schaller 1972); and lion predation on zebras can be intense enough to cause a reduction in the size of group and perhaps even of whole populations (Smuts 1976a, Sinclair and Norton-Griffiths 1982). Why mesic habitat equids are phylogenetically constrained to tend is unclear.

In view of the feeding competition in the large group, it is surprising that the females did not split up into independent harems in the winter. The grouping of the harems may have been imposed on the females by the stallions—this hypothesis implies an intersexual conflict of interests such as has been found in many mating systems (Clutton-Brock 1989).

Chapter 7   Reproduction and Growth

Because of the ground which is dismayed, since there is no rain on the land . . . .
Even the hind in the field forsakes her new born calf because there is no grass.

Jeremiah 14: 5,6

In most ecosystems, the nutrients available to large herbivores are inadequate for reproduction and growth for part of the year (e.g., Jarman and Sinclair 1979). The most costly of these processes, lactation, is usually timed to coincide with the growing season of the plants. Animals often lay down nutrient reserves, which they draw on for survival during the nongrowing season (e.g., Caughley 1970b, Mitchell et al. 1976). During the growing season, reproductive processes such as lactation may therefore compete for nutrients with the animals' body reserves unless rates of nutrient assimilation are high. In individuals that breed before they have stopped growing, which is often the case in large herbivores, there will be added nutrient requirements for the mother's own growth.

The chemistry of natural forages is such that rates of nutrient extraction are often inadequate for all these requirements at the same time; breeding females must then choose which of these requirements will receive less than optimal amounts. Evidence that females are faced with this kind of "decision," without implying that they perceive it consciously, comes from comparisons between the fat reserves of breeding and nonbreeding females. Nonbreeders have, on average, greater nutrient reserves than breeders (e.g., Mitchell et al. 1976), and nonbreeders have a better chance

Table 1. Some Published Rates of Reproduction in Horses

Location	Foals/mare (%)	Source
France, domestic	>3 yrs old, 88 <3 yrs old, 77	Martin–Rosset and Palmer (1977)
United States, feral	c.50	Feist and McCullough (1976)
United States, feral	61	Keiper (1979)
United States, loosely managed	74	Keiper (1979)
Great Britain, loosely managed	46	Tyler (1972)

of surviving the next non-growing season (Clutton-Brock et al. 1982, p. 77; Laurie and Brown 1990). The way in which a female allocates resources to her own growth, her body reserves, and the milk for her offspring will therefore affect her own growth and survival, as well as the growth of her offspring.

Several theories of life history strategy predict that natural selection will favor individuals which breed early in populations at low densities. When nutrients are abundant, the costs of breeding are low, and the benefits (in terms of lifetime reproductive success) are high. In high-density populations, selection should favor a "k-type" strategy in which the females start breeding at a later age and invest more in individual offspring (e.g., Kozlowski and Weigert 1986).

There is a large body of evidence from studies of both domestic (Allden 1970) and wild species (Albon et al. 1983) that in populations at high densities, females, on average, breed later and less frequently. What has received less attention, though, is the variance among individuals. It is clearly crucial to an understanding of population dynamics to know whether resources are shared evenly or unevenly among the females. An even distribution will usually lead to a higher average rate of reproduction than an uneven one.

In this study, the food resources were of inadequate quality for lactation and growth in the winter, regardless of the abundance of the food. The population density increased fivefold, peaking in 1981. Reproductive rates of horses are reportedly variable and often low; see Table 1. I therefore predicted that as the density of this herd rose, so the reproductive rate of the females would fall. The results of the study of their diets showed that resources were not being shared evenly (see Chapter 4, Fig. 10) and that older, dominant individuals had nutritionally better diets than other horses. I therefore predicted that dominant mares would raise more, or heavier, offspring than subordinate ones.

Table 2. Criteria Used for the Visual Condition Index

Condition class	Critieria
1. Excellent	"Blocky" appearance, skeletal structures not perceptible
2. Normal	Ribs and pelvic bones covered
3. Moderate	Ribs perceptible, pelvic bones covered
4. Thin	Ribs visible not prominent, pelvic bones slightly covered
5. Very thin	Ribs and pelvic bones prominent
6. Emaciated	Ribs and pelvic bones project very prominently; abdomen often has a "pinched" appearance; animal may be weak

Note: Data based on Lowman et al. 1976, Pollock 1980.

In this chapter, I first present data on changes in female body condition, the food available, growth rate, and fecundity as density increased. I then examine how females allocated their limited nutrients between their own growth and that of their foals. Finally, the performance of dominant animals is compared with that of subordinates. Methodological details are provided in Appendix 9.

Seasonal and Year-to-Year Changes in Condition

Body condition was evaluated according to the criteria in Table 2, with the precautions proposed by Evans (1977). The same basic seasonal pattern of changes in condition of breeding mares was observed each year; see Fig. 1. Their condition was best in autumn, worst in late winter.

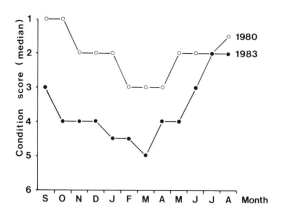

Fig. 1. Seasonal changes in the median score of breeding mares on the visual index of body condition (1 = excellent, 6 = emaciated) in a good and a bad year.

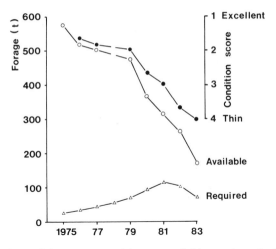

Fig. 2. Estimations of the amounts of forage available on the principal pasture in January of each year and the amount required by the horses for January–March, and the mean condition score of breeding mares in January.

There were also year-to-year variations in condition: the animals became thinner from 1980 to 1983. The mean condition score in January is shown in Fig. 2, together with the estimated amount of food available at that time, and the estimated food requirement for January, February, and March (see Appendix 9). There was a close correlation between their condition score and the available food ($r_s = 1.0$, $n = 7$, $p < 0.05$), although the estimate of the amount of food required was never more than 50% of what was available.

The mean body weight of mares >6 years of age in September and March is shown for each ecological year in Fig. 3. March weights of gravid females were liveweight less the conceptus and mammary gland,

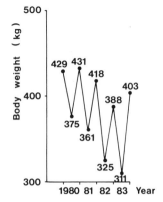

Fig. 3. Mean body weights of mares ages 7 years and older in September and March of each ecological year (September–August), less the weight of the conceptus.

estimated using the method of Martin-Rosset and Doreau (1984a). The years of population decline, 1982 (September 81–August 82) and 1983 were clearly more difficult nutritionally than 1980 and 1981: The autumn weight declined by 10% in 1982 and the spring weight by 17% between 1980 and 1983. The amplitude of the seasonal change, 14% in 1980, increased to 24% in 1983.

The causes of the stress of 1982–1983 were presumably the reduction in the quality of food, especially marked in spring (see Chapter 4, Fig. 7), and the quantity of food, especially in winter; see Fig. 2.

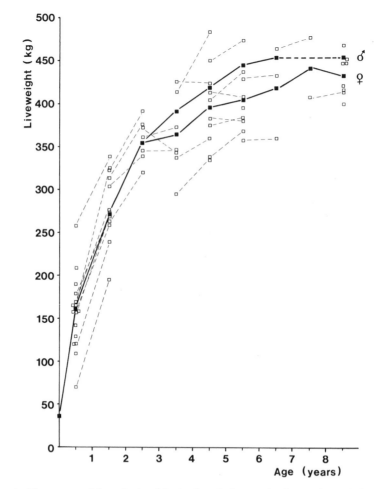

Fig. 4. The age–weight relationship in female horses in September of the years 1979 and 1980. The dotted lines join the values from individual females, and the solid line the mean values. The mean values for males are also shown.

Growth of Female Horses

The weights of female horses in September of the years 1979–1980 are shown in Fig. 4, together with the mean values for males. The value for males aged 8–15 years is taken from 1983, as there were insufficient data from the previous years.

The growth curve shows that females grew rapidly until 2 years, then more slowly until they were aged 7 years, even in these nutritionally good years. Males grew at a decelerating rate until 6 years. Sexual dimorphism in weight was very slight ($<5\%$), but males weighed on average more than females from age 2 years onward. This pattern of growth (monomolecular) is typical of horses, but the relatively late age at which the asymptote was reached indicates that the animals' nutrition was suboptimal (Martin-Rosset 1983), even for the adult males, whose median condition score in March, at the end of winter, was never below 2 (normal), and no male was ever in worse condition than 4 (thin). The slow growth observed in this herd was, however, comparable with data on zebras, which also reach adult weight at 6–9 years (Joubert 1974, Smuts 1975b).

The nutritional stress on the females was considerably greater than on the males. The slowing down of growth in females was caused by reproduction: nonbreeders between 2 and 5 years of age gained 30.7 ± 16.2 kg (95% c.l.) more than breeders in summer. The nutritional status of females was therefore inadequate to allow both a rapid rate of growth and reproduction.

Timing of Births

Primiparae foaled on average almost a month later than multiparae (27.2 ± 13.0 days, 95% c.l.). The median birth date of foals of multiparae is given in Fig. 5 for the 10 years before the study and for each year from 1974–1983. The standard deviation of the mean date is also shown. During the first 3 years, the median birth date advanced each year. By 1976, it was in February. This was possible because the mean gestation period of mares of light-horse breeds is 335 days (Evans et al. 1977); if the mares came into estrus in a "foal heat" c. 7 days after parturition, then they would, on average, foal again 342 days later. After 1976, the median birth date became later; in 1980, it was April 23rd, close to the date observed in the 10 years before the study. There was little change in the next 3 years. The median interval between successive births, 348 days in 1974–1975, increased rapidly as maternal condition declined ($r_s = 0.96$, $N = 9$, $p < 0.05$), reaching 373 days in 1983. A condition score of 3 in January corresponded to a median interval of 1 year.

The births were spread over the months November–August, and there was a significant tendency for the births to be more scattered in years when

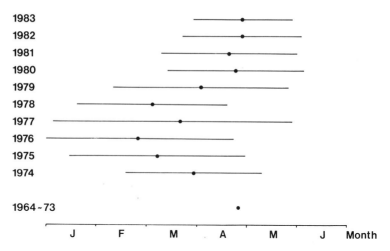

Fig. 5. The median date of birth ± 1 *SD*. As birth date increased, the standard deviation declined ($r = 0.65$, $p < 0.05$).

the foals were born early (Fig. 5), but even in 1983, the standard deviation was 30 days.

Camargue horses therefore normally foal around the last week of April, with two thirds of the births in April and May. Superabundant food in the first years of the study allowed them to foal at intervals of less than a year, and in a less strictly defined season. In the springs of 1978, 1979, conceptions were delayed, so in 1980, births once again centered on late April.

The coincidence of the normal median birth date and the peak of diet quality (see Chapter 4, Fig. 6) was striking; the intake of digestible protein was very much higher in early May than at other times (Chapter 4, Table 9).

Fecundity

Among the older mares (>7 yr), *fecundity* (foals born per ♀) was close to 100 (96%, $n = 73$) in the years 1975–1980; see Fig. 6. The only mares that did not foal each year were one who became permanently barren at 3 and one that was 23 years of age. The few fillies (2–3 years, $n = 4$) in the first 2 years all foaled, then from 1976–1981, the fecundity of fillies ranged between 55 and 75%. An exception here was 1977, when only one out of eight foaled. Fillies show reduced levels of sexual behavior with the dominant stallion of their natal group (see Chapter VI, Table 3). In spring 1976, the dominant stallion drove out all the younger males, so that by July, he was the only male more than 1 year of age in the group. The lack of other stallions may explain the failure of these fillies to conceive.

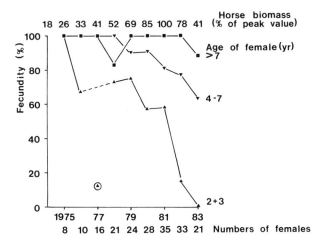

Fig. 6. Fecundity of adult and younger females in each year.

During the last 3 years, however, while the fecundity of the mature mares remained at 95% ($n = 22$), the mean fecundity of the younger mares (4–7 years) fell, reaching 63% ($n = 8$) in 1983. Only two fillies (2–3 years, $n = 18$) bred in the last 2 years.

Female fecundity therefore showed little or no response to the increase in density until 1981, when the peak density of 0.26 horses, or 81 kg live weight (LW) per ha was reached. The fecundity of *young* females then declined sharply.

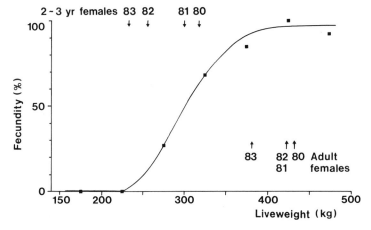

Fig. 7. The relationship between liveweight in September (all females, 1979–1982) and fecundity in the following spring. The arrows show the median weight of 2–3 year-old and adult mares in each of the ecological years 1980–1983.

Fecundity was closely related to body weight; see Fig. 7. No females weighing less than 250 kg in September foaled in the following spring; on average, one in two females weighing 300 kg foaled, and 95% of females weighing >400 kg foaled (the barren female was excluded from this analysis).

Autumn weights of adult females showed little change until the last year (1983), and even then, the median weight (381 kg) still corresponded with a fecundity of >90%; see Fig. 7.

Autumn weights of younger females were much more sensitive to shortages of food (see Fig. 7) and declined by 27% between 1980 and 1983. This decline in weight presumably caused the sharp decline in fecundity shown for subadult females in Fig. 6.

Mortality of Young Horses

Neonatal mortality (0–2 days) was higher in foals of 2- to 3-year-olds (21%, $n = 34$) than in foals of older females (7%, $n = 113$, $\chi^2 = 8.8$, $df = 1$, $p < 0.01$, female ages 4+ years 1973–1983). Among these older females the number of surviving foals per female was 1.0 in the years 1975–1977 ($n = 21$, Fig. 8) and then declined to 0.69 in 1983 ($n = 16$). The relatively high rate of neonatal mortality in 1978 was probably caused by the social instability at that time (see Chapter 6). The rate of production of living foals therefore began to decline when the biomass of the herd was at about a half of its peak value (cf. Fig. 6). These failures to breed successfully

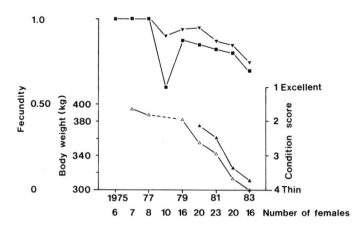

Fig. 8. The proportion of mares >3 years of age foaling (inversed filled triangles) and whose foals survived the first 48 hours (filled squares). The body weight in March (filled triangles) and mean condition scores in January of the same years are also given (open triangles).

($n = 13$) were not significantly more common in subordinate than in dominant mares, (13% vs. 14% per year, $n = 54$ vs. 43, years 1978–1983).

Infant killing is known in many of the mammals with harem-based social systems (Hausfater and Hrdy 1984). It has been observed in herds where the owners simulate natural takeovers by switching the stallions between years (Duncan 1982). In this herd though there were no takeovers of groups, in none of the cases where mares switched family units did were either infant killing or forced copulations leading to abortion (cf. Berger 1983a) observed. The high level of neonatal mortality in 1978 (see Fig. 8) was, however, associated with a time of social instability (see Chapter VI), and fighting between the stallions was frequent. Some of the foals died of abandonment, others were drowned. It seems likely that the social instability increased the probability of these events.

Foals that survived to 2 days suffered no further mortality (with one exception, due to a broken leg) until they were weaned. This occurred in winter, and 10–20% of the foals died or lost condition to the point where they had to be removed in each of the years 1980–1983. Males and females survived equally well to 1 year of age (sex ratio at birth = 49% ♂, $n = 137$ at 1 year 45% ♂, $n = 107$).

Lactation and Growth

Between 1975 and 1983, the duration of lactation increased, while the dry period decreased in length; see Table 3. Mothers therefore spent a greater part of the year lactating in the nutritionally difficult years than in easier ones. There was no difference in the length of lactation between mares with male and mares with female offspring.

Nonpregnant mares lactated for nearly 28 weeks longer than pregnant ones on average ($\bar{X} = 71$ weeks; $t = 15.1$, $df = 6$, $p < 0.001$). Robbins (1983) states that the milk received by young deer at the end of lactation is not nutritionally significant. This was not so for the horses: the yearlings of the few barren mares were visibly larger than their peers, and their weight gains were greater both in winter and even during summer when their food was abundant; see Table 4. There was no doubt in this herd that the milk the young horses received in their first winter was crucial for their growth, and survival.

Growth Rates of Foals in Their First Summer

Effect of Mother's Age

The three foals born very early (January) were notably lighter in September (-35 kg) than those born during the normal birth period. Foals born earlier than the mean birth dates less 2 SD ($n = 6$) were therefore omitted

Table 3. The Duration (Weeks) of Lactation and the Dry Period of Multiparae

	1975	1976	1977	1978	1979	1980	1981	1982
Lactation								
median	35	34	$38\frac{1}{2}$	36	$40\frac{1}{2}$	$38\frac{1}{2}$	43	45
range	20–40	27–36	28–42	34–39	35–49	27–46	33–49	41–62
Dry period								
median	15	$16\frac{1}{2}$	$14\frac{1}{2}$	16	14	15	12	9
n	7	7	4	5	16	17	13	6

Note: Data reported for each year of the study, with male and female foals combined. Lactation increased ($r_s = 0.91$, $p < 0.05$) during the 8 years, and the dry period decreased in the last two.

Table 4. The Effect of Suckling on Growth of Foals in 1983

Period	Weight changes (mean)		
	December–March (g/day)	April–September (g/day)	September–September (kg)
Suckling foals	+106	700	+151
Weaned foals	−54	493	+95
n	5, 5	4, 3	3, 3

Note: Those that were weaned in the period December–March gained less weight than those that continued suckling both in winter ($t = 6.0$, $df = 8$, $p < 0.001$) and in summer ($t = 3.62$, $df = 5$, $p < 0.02$).

Table 5. Analysis of Variance of the Effect of Year of Birth and Mother's Age on the Weight/Age Relationship for Foals in September of Their First Year

	df	F	p
Variable			
Year	4, 37	9.7	<0.001
Mother's age	1, 37	7.3	<0.01
Interaction	3, 37	1.4	n.s.
Deviations (kg)			
Year			
1979		0.77	
1980		18.5	
1981		−0.78	
1982		−11.8	
1983		−10.0	
Mother's age			
> 6 years		4.6	
5, 6 years		−5.6	

Note: Deviations are adjusted for independent variables.

from this analysis. A foal's gender had no effect on its weight in its first autumn, ($F_{1.28} = 0.0$, $p = 0.99$): the sexes have therefore been combined.

The year of birth had a strong effect (see Table 5): the heaviest foals of 1980 weighed 30 kg (c. 20%) more than those of 1982–1983. This analysis also showed that the age of the foal's mother had an important effect: foals of mature females weighed 10 kg more than foals of mothers aged 5–6 years.

The only years for which there were adequate numbers of mares in all age classes were 1979 and 1980. The analysis was therefore restricted to these years (see Table 6), and it was found that the weight of the foals increased regularly with mother's age. The regressions for each age class, with the effect of year removed, are given in Fig. 9. Foals of 2- to 3-year-

Table 6. Analysis of the Effect of Mother's Age and Year on the Weight/Age Relationship for Foals in Autumn

	df	F	p
Variable			
Mother's age	3, 25	14.0	<0.001
Year	1, 25	4.4	<0.05
Interactions	3, 25	1.55	n.s.
Deviations (kg)			
Mother's age			
>6 years		14.4	
5, 6 years		5.9	
4 years		−7.4	
2, 3 years		−17.4	
Year			
1979		−4.3	
1980		3.2	

Note: Deviations are adjusted for independent variables.

old fillies were on average 30 kg lighter than foals of adult females at a given age.

The most obvious explanation for this was that larger mothers produced heavier foals. The regression of foal deviation on mother's weight in September across all ages was significant in each year, but when the effect of mother's age was removed, the regressions were weak and nonsignificant. Mother's weight per se was not the main determinant of foal weight.

During summer, the mares used nutrients for building up body reserves for use in winter (see Fig. 3), as well as for producing milk. The younger mares were also growing (see Fig. 4). The fact that these younger mares produced lighter foals suggested that they grew at the expense of their foals. If this was so, then

1. Younger mares should put on more weight in summer than adults
2. There should be a negative correlation between the weight of the foal and its mother's summer weight gain.

These predictions were tested on the data for 1980 and 1981: The other years had few young females. Adult mares were given a nominal age of 8 years, as there was no consistent trend thereafter. In each year, the mother's summer weight gain decreased with age (see Table 7), the younger mares gaining 15–20 kg more than the adults.

To test the second prediction, the effect of a foal's age on its weight was removed by calculating the deviation of each from the general regression of weight in autumn on the foal's age. Relatively heavy foals had a positive

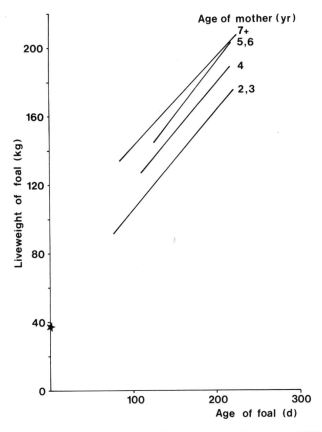

Fig. 9. The age/weight relationship for foals whose mothers are of different ages. The mean birth weight is shown by a star (38 kg, range 28–48 kg).

Table 7. The Summer Weight Gain of Breeding Mares, as a Function of Their Own Age and the Correlation Coefficients of Summer Gain with Age

Year	Maternal Age (yr)				Correlation coefficient	
	2, 3	4	5, 6	>6	r	n
1980	70	64	59	57	−0.63**	16
1981	77	65	71	57	−0.75**	16

**$p < 0.01$.

deviation, light ones a negative one. The regression of the foal's deviation on the mother's summer gain was then calculated. In each year, mothers which gained much weight themselves produced lighter foals ($r = -0.64$, -0.53; $n = 16$, 16; $p < 0.01$, 0.05), and the slope for both years combined was close to unity ($b = -1.02$).

Effect of Mother's Rank

The effect of social status was examined by pairwise comparisons of the offspring of dominant and subordinate mares. These comparisons were restricted to mothers in the same year, family, and age class. The results show that foals of dominant mares were 7.9 ± 3.0 kg heavier per unit difference in rank ($t = 3.6$, $df = 24$, $p < 0.01$). This difference was unlikely to be a consequence of the slight ($+2.2\%$ of liveweight) and nonsignificant ($t = 1.1$, $p > 0.2$) size difference between dominant and subordinate mothers.

A measure of resources available to a mare is the sum of her own gain plus her foal's deviation from the general regression. A mare which put on a lot of weight and which had a heavy foal would score high on this measure, and vice versa. The same paired comparison was made as above, and the dominant mares were found to score higher than subordinates ($+9.9 \pm 5.1$ kg 95% c.l.) on this measure.

Effect of a Mother's Age and Rank on the Age at First Foaling of Her Daughters

For fillies of the first 3 years, 1974–1976, there was no significant correlation of mother's age or rank with the age at first foaling of her offspring ($r = 0.08$, 0.06, $n = 12$, 12).

In the next 3 years, mother's age accounted for 54% of the variance in the age at first foaling (24–48 months, $b = -2.45$ months $n = 14$, $p < 0.01$). When this effect was removed, mother's rank accounted for a further 39% (24–42 months, $b = 1.34$, $n = 14$, $p < 0.05$); see Fig. 10.

Population Growth

The high fecundity allowed the Camargue population to grow at a rate of c. 30%/yr ($\lambda = 1.30$, $r = 0.24$) when it was not managed; see Table 8. This confirms the view that "it appears that λ values >1.20 are definitely possible in some (equid) populations" (Wolfe et al. 1989, p. 923).

Such a rate of increase is very high for an ungulate. The Granite Range mustangs' increased at 20% ($r = 0.188$, Berger 1986) per year, while the wildebeest and buffalo of the Serengeti increased by c. 10%/yr after the disappearance of rinderpest (Sinclair 1979). The high rate of increase in the Camargue was permitted by the very high annual survival rates of adults

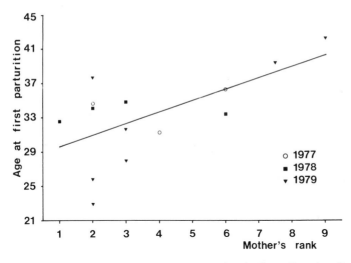

Fig. 10. The effect of mother's rank on the age at first foaling of her daughter, with the effect of mother's age removed ($r = 0.622$; $y = 1.37x - 4.77$; $p < 0.05$).

Table 8. Rate of Increase in the Camargue Herd during the Time It Was Not Managed

Year	N	Increase (%)
1974	14	—
1975	20	43
1976	26	30
1977	35	35
1978	43	23
1979	56	30
1980	75	34

Note: N = total number of horses before foaling began.

(100%, >6 years) and young horses (98%, 1–6 years) in this predator-free environment.

The removals of very thin animals stopped the increase but biased the adult sex ratio (54% male, $n = 61$ in 1981). A similar pattern was found in a food-limited feral horse population (Sable Island, 56% male, from data in Welsh 1975).

Such male-biased sex ratios are unusual in natural equid populations where survival in males is generally lower than in females, apparently due to a combination of predation and male–male competition (Berger 1983b).

General Discussion

Effect of Density on the Timing and Rates of Reproduction

When the horses' density was very low, breeding was less strictly seasonal. However, when the density of animals was about a half of the peak value, and before condition declined notably (1978–1979), conceptions were delayed, so that by 1980, births were centered on the last week in April. The considerable increase in density which occurred after 1979 had no further effect. The birth peak then coincided exactly with the time when the quality of the mothers' diet was maximal, which presumably contributed to high milk production. Early born foals grew more slowly than others, both in this study and in thoroughbreds (Hintz et al. 1979).

The fecundity of young females, as in many ungulate populations (e.g., Sinclair 1979, Fig. 4.8) and as predicted above, was very sensitive to food shortages. In this study fecundity of 2- to 3-year-olds fell from 100 to 0% (see Fig. 6), presumably because reduced food supplies in the later years caused slower growth rates in the young females.

The fecundity of adult females (7+ years), on the other hand, was close to 100% and failed to respond as predicted to the nutritional stress of the last years. The fecundity of Granite Range mares increased to 0.92 at 7 years but declined to around 0.75 after 10 years (Berger 1986). The fecundity of the Camargue mares is the highest which has been reported in any unmanaged equid population (e.g. Table 1 and Welsh 1975, Smuts 1976b, Seal and Plotka 1983, Berger 1986, Wolfe et al. 1989). Ninety-five percent fecundity is unusual among free-ranging ungulates (e.g., buffalo, 50–66%, Grimsdell 1973; lactating red deer, 44–67%, Lowe 1969). According to the fecundity–weight relationship (see Fig. 7), Camargue mares would have to lose 25% of their normal autumn weight (425 kg) before fecundity fell to 66%. This implies that fecundity in adults would fall only in years of major die-offs. Fecundity that is so high and so insensitive to undernutrition may be a peculiarity of the Camargue breed among horses (see also van Niekerk and van Heerden 1972); a similar pattern is found in humans (Bongaarts 1980).

The very high fecundity of Camargue horses is difficult to explain. Perhaps some aspect of artificial selection on this breed has led to increased fecundity: this is unlikely, for the genetic component of the processes controlling fecundity in large mammals is notoriously slight. The only alternative explanation is that the measures of fecundity in feral horses include neonatal mortality. This would not explain the low level in zebras, which was measured on a sample of killed mares.

Such a high fecundity rate was certainly the main cause of the extremely poor condition and the survival of the females in both food-limited horse populations so far studied (Camargue and Sable Island). In such circumstances, the cost of reproduction in females is clearly higher than in males.

Allocation of Resources between Maternal and Offspring Growth

Female farm animals allocate their resources differently according to their age: young, growing females produce lighter calves than mature females (Bourdon and Brinks 1982). In a study of thoroughbred horses, the fastest-growing foals were those of 8- to 11-year-old mothers (Hintz et al. 1979).

In the years 1980–1981, when the food of this herd was not superabundant, females could not simultaneously reproduce and grow fast. Young females (<6 years) which were still growing put on more weight in the summers than did mature females, and they produced lighter foals as a consequence. They therefore compromised, growing more slowly than nonbreeders, and investing less in their foals than did adult females.

Adult females had stopped growing, but all built up their nutrient reserves for winter. Between 1980 and 1983, they put on progressively more weight themselves, and produced progressively lighter foals in autumn (see Fig. 3 and Table 5). They extended lactation, using these reserves, in the last years, so that their foals received some milk throughout their first winter, which would have helped them to survive—without milk, they lost 50 g/day.

Effects of Maternal Rank

Dominant mares had more nutrients available to them than subordinates (mother's gain + foal deviation) as a result of the better quality diets of the dominants (see Chapter 4, Fig. 10). Contrary to the preceding prediction, dominant mothers produced no more foals than subordinate ones, but their foals were heavier.

What effect did these differences in maternal investment have on the reproduction of the offspring? The sons of dominant mares have higher reproductive success than the sons of subordinates, but apparently not because they are heavier (Feh 1990). Daughters of dominants foaled earlier (by a year) than daughters of low-ranking mares. It is probable that these differences in age at first foaling were direct consequences of the differences in the growth rates of the daughters. Because age at first reproduction is a key component of lifetime reproductive success (LRS) in many species (see papers in Clutton-Brock 1988), it is likely that the LRS of daughters of dominants was enhanced.

More generally, there is considerable evidence that dominant male mammals, especially in carnivores and ungulates, have higher rates of copulation, and there is some evidence that this leads to higher LRS than for subordinate males (see Dewsbury 1982). In females, the evidence that measures of reproductive success are related to rank is less strong (but see Robinson 1982, Silk 1987). One explanation for the paucity of positive results is that dominance may affect the survival of offspring only periodically (Fedigan 1983). The results of this study show that maternal dominance had a strong effect on the growth rate of offspring, even if it did not affect offspring number or survival. An important aspect of these results is that the daughters of dominant mares bred almost a year earlier than daughters of subordinates when the population was at a high density, but not when density was low. Thus, even the effect of maternal dominance on the growth rate (quality) of their offspring was a periodic phenomenon and was expressed only when resources were scarce.

Low-ranking females therefore produced as many foals as dominants; and reproduction at 2 years of age had no effect on the probability of foaling the next year. The reproductive rate of adult horses therefore showed little response to increasing density and scarcity of resources. Females delayed their first breeding attempt, but there was little change in the fecundity of older mares and no change in adults. They invested smaller amounts of nutrients in their foals, more in themselves. Their reproductive strategy was to maximize the *number* of offspring, not their quality as measured by weight.

It is tempting to model this system in order to evaluate the adaptiveness of this strategy in a species where weight has no effect on LRS—in males, at least. To do this, one would need both a measure of the effect of weight at weaning on LRS, and a measure of the effect of lactation on the mother's probability of dying in winter. The nature of this study meant that such data are not available: a similar study will need to be carried out on wild equids to explore this area further.

Population Regulating Factors

This herd was regulated by removing thin animals after the herd reached the food ceiling. At this stage, the decrease in fecundity, and neonatal mortality, were slight. Without management, the population would not have been regulated by any factor other than increased mortality among weaned horses due to starvation: in particular there was no suggestion that these horses had an intrinsic mechanism of population regulation. The generality of this conclusion is discussed in the last chapter.

Section D   Equids and Their Habitats

Chapter 8   The Impact of Grazing on the Plants
and Animals of the Camargue

En hiver c'est un marais; en été c'est un desert. Le marais est peuplé d'innom-
brables oiseaux aquatiques; le desert par des nuées de moustiques."[a]

Pader 1890

Wetlands have traditionally provided people with food from fisheries and
wildfowling. In the fenlands of eastern England in 1724, the numbers of
wildfowl were "incredible, and the accounts the country people give of the
numbers they sometimes take, are such that one scarce dares report it!"
(Darby 1983, page 141). These wetlands are now farmland.

Wetlands have also provided lush grazing during the warm seasons.
Most wetlands in Europe have been grazed by livestock since the disappearance of the wild herbivores. The famous "pré-salé" lamb comes
from the salt marshes of northern France, and the same Lincolnshire fenlands contained "prodigious numbers of large sheep and also oxen of the
largest size" (Darby 1983, page 138).

Much of the wetlands of northern Europe have disappeared through
drainage and reclamation in the past 300 years, but large areas remain,
especially in the Waddensee and western France. Many of these wetlands
are managed for wildfowling and some for conservation of their populations of wild plants and animals. In recent years, there has been a decline in
the traditional use of the marshes for grazing largely for economic reasons:
this reduction of grazing has led to profound changes in the ecology of

---

[a]In winter, it's a marsh; in summer, a desert. The marsh hosts innumerable waterbirds; the desert, clouds of mosquitoes.

the wetlands. Studies in northern Europe and North America have documented important differences between ungrazed and grazed marshes (e.g., Beeftink 1965). Others have described the pattern of succession which follows abandonment of wetlands. In all cases, increases in plant height and cover precede a decline in plant species richness and diversity (α- and β-diversity; see Ranwell 1972, Bakker and Ruyter 1981). This process occurs rather slowly in salt marshes but fast in the more productive freshwater marshes. The formation of peat coupled with sedimentation leads ultimately to the formation of woodland. The great majority of wetland plants and animals, of course, disappear (Gordon et al. in press).

The role of grazing in deflecting or arresting this process is well understood in northern Europe and North America, and grazing animals are used as management tools on many reserves, both to deflect succession and to create good grazing conditions for graminivorous waterfowl such as geese and wigeon (e.g. Kaiser et al. 1979, Thomas et al. 1981, Gordon et al. 1990).

In the Camargue, peat formation is minimal so it could be expected that successional changes would be much slower than in northern Europe, and the impact of grazing mammals correspondingly less important. However the ecological role of grazers in Mediterranean wetlands is considerable: it is well known to wetland managers but virtually undocumented.

In this study, the horse herd was allowed to increase with minimal management. Their impact on the vegetation was studied in order to document the impact of a large mammal on Mediterranean vegetation and to provide early warning of overgrazing, which might call for a change of management. On the basis of the diet study, the first effects could be expected in the wetlands where the horses' preferred food plant, *Phragmites*, grew. The absence of other large mammal grazers provided a remarkable opportunity to evaluate the effect of an equid on a wetland ecosystem.

In several other studies, the impact of large mammals has been assessed by excluding grazing from a set of study plots and following succession in the ungrazed and grazed swards (see Watt 1962 for a study of rabbit grazing; for sheep, see Rawes 1981, 1983; for cattle, see Bakker 1978, 1985). The technique is valuable in systems where the animals have a strong impact, but as Bassett (1980) has pointed out, the pressure of free-ranging animals on the plots cannot be controlled, and the small size of most exclosures can lead to anomalies. For instance, in this study, the horses had so strong an impact on the emergent plants of the marshes that the coypu used the exclosures fenced against horses preferentially, and wild boar developed such a taste for the roots of the plants in protected plots that it proved very difficult to exclude them for more than a few years. These effects limit the usefulness of exclosures in the long term, but they proved invaluable in the short term.

A system of exclosures was therefore designed by Philip Bassett, which excluded horses and the smaller mammals from study plots (but allowed in invertebrate and small rodents such as *Microtus*). Next to these were sited plots from which only horses were excluded. It was not feasible to obtain plots grazed by horses alone. As rabbits did not graze the deep marshes, and coypu rarely left them, the effects of the three larger herbivores were studied in the following years by Jacques Jardel, by Jean-Marc and Marie-Christine d'Herbès, as well as by P. Bassett.

In addition, a set of smaller exclosures was set up each year to measure the peak standing crop in the marshes, a measure of above-ground net primary production (ANPP, Linthurst and Reimold 1978), and in 1982, the method of Milner and Hughes (1968) was used by J.-M. d'Herbès to measure primary production under grazing in the marshes and grasslands; see Appendixes 4, 10 for details of the methods used.

Food Intake of the Horses

The intake was calculated in two different ways: first, by calculating the mean biomass of the horses for each year and multiplying this by 4.2%, the median daily intake (see Chapter 4). The second method was based on the published energy requirements of horses; the results corresponded closely. The horses' intake peaked at c. 400 tons in 1981; see Fig. 1.

The annual net aerial primary production was estimated to range between 2000 and 3500 tons of dry matter (see Table 1), so the offtake increased from 3 to 5% of production in the early years to about 20% in 1981. These levels of intake by horses are similar to those reported in other wetlands (11%, Keiper 1981; 13 %, Welsh 1975; see also Eline and Keiper

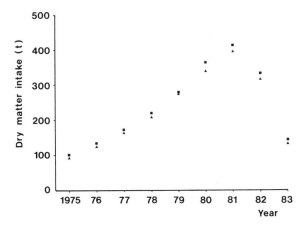

Fig. 1. Annual dry matter intake (DMI) of the horses estimated both from their energy requirements (triangles) and from their biomass, assuming a daily intake of 4.2% of their liveweight (squares; see Appendix 10, p. 256 for details).

Table 1. A Rough Estimate of Annual above-ground Net Primary Production (ANPP) for the Whole Pasture

		ANPP			
		(t/ha/yr)		(t/yr)	
Facet	Area (ha)	High	Low	High	Low
Deep marsh	26	15	11	390	286
Shallow marsh	25	15	11	375	275
Fallow fields	21	14	11	294	231
Enganes–sansouïre	179	10	5	1790	895
Halophyte grassland– coarse grassland	83	6	2	498	166
Total				3347	1853

Note: "High" and "low" are estimates derived from the upper and lower limits of the range of production values (see Chapter 2, Table 4).

1979) and much lower than the levels reported from a temperate forest ecosystem, the New Forest (c. 100%, Putman 1986).

The offtake estimated for the rabbits (intake + wastage) in 1983 was 124 tons (see Chapter 2, Table 5), concentrated in the higher facets, especially halophyte grassland (Rogers 1981).

Impact of Grazing on Herbaceous Plants

The Marshes

The plots protected from horses (X, Y) in the shallow marsh site maintained a high degree of plant cover (until invaded by wild boar in 1981) and height increased (see Fig. 2, Graphs a and b). In the plots grazed by horses, on the other hand, cover and height declined sharply after 1978.

Only two plant species occurred in appreciable densities here, *Scirpus maritimus* and the stoloniferous grass *Aeluropus littoralis*. *A. littoralis* tended to decline through the years; *S. maritimus* on the other hand tended to increase in all plots (see Fig. 3), except in 1982, when the frequency was lower in the grazed plots than in any year after 1976.

Protection from grazing therefore allowed the tallest plant (*S. maritimus*) to increase in frequency, height, and cover. Under the increasing grazing pressure, height and cover declined, but the frequency of *S. maritimus* was affected by grazing only in 1982.

In the deep marsh site, the changes were more marked than in any other. By the end of the first growing season, the shoots were taller inside

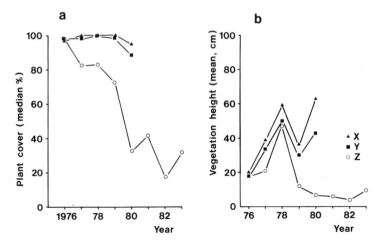

Fig. 2. (a) Plant cover in the three treatments (solid triangles [X] = ungrazed; solid squares [Y] = grazed by rabbits; open circles [Z] = grazed by rabbits and horses) of the shallow marsh site; wild boar rooted all so-called protected plots in 1981; (b) the height of the vegetation in the three treatments of the shallow marsh site (treatments as in Graph a).

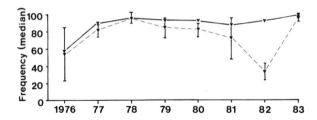

Fig. 3. The frequency of *Scirpus maritimus* in the shallow marsh site tended to increase in both the ungrazed plots (solid line) and the plots open to horses (dotted line). The frequency in the grazed plots in 1982 was, however, significantly lower than in any other year 1977–1983 (the bars are 95% c.l.).

than outside the annual exclosures (mean height 101 vs. 64 cm, $t = 4.1$, df = 5, $p < 0.001$); density was not significantly different (121, 125 shoots/ m²). The height of the vegetation increased greatly in the plots permanently protected against horses (X and Y), and a dense reedbed quickly appeared; see Fig. 4 and photographic plates. The frequency of *Phragmites* increased from 20 to 100% in the protected plots but declined outside; see Fig. 5. Density in the whole of the grazed marsh declined by 60% between 1975 and 1976 (see Fig. 6) and had fallen to 0.26 shoots/m² by 1983. The

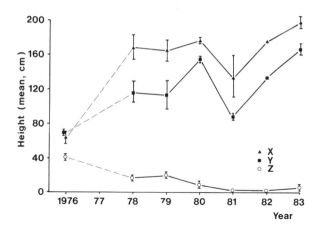

Fig. 4. Mean height (± 1 SD) of the vegetation in the plots of the three treatments (X = ungrazed; Y = grazed by coypu; Z = grazed by coypu and horses) in the deep marsh site.

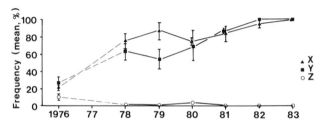

Fig. 5. Mean frequency of *Phragmites* (± 1 SD) in the vegetation in the plots of the three treatments in the deep marsh site (symbols as in Fig. 4).

frequency of *S. maritimus* in the ungrazed plots remained between 70 and 100%, while in the plots grazed by the horses, it fell gently between 1979 and 1982, then sharply to less than 20% in 1983; see Fig. 7.

The horses therefore had a very strong effect on *Phragmites* even when their food intake was a only quarter of its peak value and only about 5% of the pasture's production. Their effect on *Scirpus* was much weaker and became strong only in 1983 after 5 years of heavy grazing.

The peak biomass of emergent plants in the annual exclosures in 1975 was 1100 g/m². This declined in the grazed marsh but increased to 2580 g/m² in the permanently protected (X) plots; see Fig. 8. Protection therefore increased the productivity of emergent plants, whereas the heavy grazing of the last years caused a decline. Production under grazing was measured in mobile exclosures in 1982 only. The estimate (125 g/m²) was significantly lower than the 405 g/m² in the annual exclosures ($p < 0.05$, sign test). In the grazed marsh, in addition to the emergent shoots of *Scirpus*

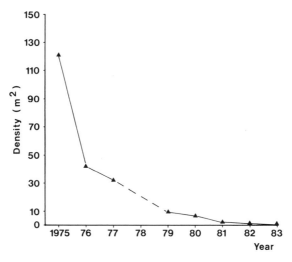

Fig. 6. The mean density of shoots of *Phragmites* in the deep marsh. There was a significant decline in four out of five sites, 1975–1976 (*t*-test, $N = 25$, $p < 0.05$).

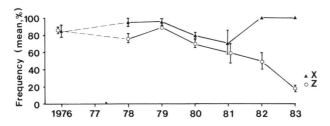

Fig. 7. Mean frequency of *Scirpus maritimus* ($\pm 1\ SD$) in the vegetation in the plots of the three treatments in the deep marsh site (symbols as in Fig. 4).

and *Phragmites*, there were also submerged plants, *Characeae* and *Ranunculus baudoti*. These plants were virtually absent from the ungrazed exclosure, presumably because of competition for light with the dense, tall reeds. Their peak biomass increased between 1975 and 1982; see Table 2. A rough estimate of ANPP net in 1982 may be obtained by adding this figure to the peak standing crop of emergents ($390\ g/m^2$). This estimated value for the primary production of the grazed marsh ($530 g/m^2$) is about one fifth of the value for the ungrazed exclosure.

In the marshes, therefore, the vegetation changed rapidly when grazing was removed: in both sites, a tall perennial monocotyledon became dominant in height and cover; see photographic plates. The grazed vegetation changed in the opposite direction, becoming shorter and more open. *Phragmites* became very rare, and *Scirpus maritimus* declined in frequency under heavy grazing, especially in the deep marsh.

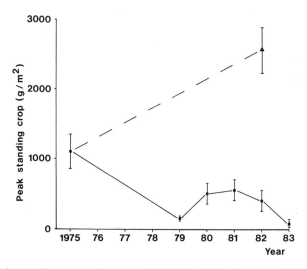

Fig. 8. Peak standing crop of emergent plants of the deep marsh in exclosures protected for the growing season (squares) and in the exclosure protected for 6 years (triangles).

Table 2. Peak Biomasses (mean ± SE in May) of Submerged Plants in the Grazed Part of the Deep Marsh

Year	Characeae (g/m²)	Ranunculus baudoti (g/m²)	Total
1975	—	—	42 ± 18
1982	241 ± 44	10 ± 4	251 ± 48
1983	284 ± 62	6 ± 1	290 ± 63

The reedbed which developed in the ungrazed plots was more productive of emergent plants than the plots protected for 1 year only, by a factor of about five. In the grazed parts, grazing caused the production of emergent plants to fall still further. The production of submerged plants was inadequate to make up this considerable drop in primary production.

The Salt Flats

The results from the enganes exclosure suggested that the grazers, particularly the horses, maintained the short, rather open vegetation of the salt flats, for cover increased with protection from both species; height was shorter when grazed; see Table 3.

Neither diversity nor the frequency of individual species changed significantly among years or treatments.

Table 3. The Cover and Height of the Vegetation in the Exclosure Site E (Enganes)

Vegetation cover	Mean (%)		Difference		Vegetation height in 1982 (mean, cm ± 95% c.l.)
	1976	1982	t_s	p	
X (ungrazed)	79	98	4.72	<0.001	33.8 ± 2.2
Y (rabbit grazing)	67	89	3.87	<0.001	22.9 ± 1.8
Z (horse + rabbit grazing)	75	67	1.24	>0.2	11.9 ± 2.6

From 1976, cover increased in the ungrazed (X) and rabbit-grazed (Y) plots but did not change significantly in plots grazed by horses and rabbits (Z). Height was shorter in the horse-grazed that in the other treatments in 1982 (t-test, $p < 0.01$).

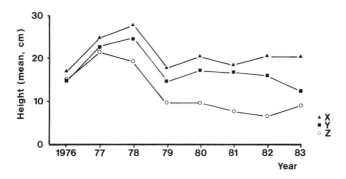

Fig. 9. The mean height of the vegetation in the three treatments of site D (symbols as in Fig. 2, Graph a).

The Grasslands

Both the cover and the height of the vegetation fluctuated between 1976 and 1982 (see Fig. 9), apparently in response to rainfall, which was heaviest in 1977 and 1978 (Chapter 2, Table 2). Within treatments, there was no significant change in the ungrazed plots; see Table 4; in site A the vegetation of the plots grazed by rabbits alone decreased significantly in both cover and height between 1976–1982; and there was an important difference between X and Y ($t = 3.39$, $p > 0.001$). In site D, there was no significant rabbit effect on either parameter.

No treatment was grazed by the horses alone, but height and cover were significantly lower in the Z treatment grazed by both species in both sites. There was therefore a horse effect independent of the rabbits. The effect of rabbits was strong (Y significantly different from X) in site A, not in site D. Both species of herbivore therefore affected the vegetation, though the rabbits' effect may have been patchy. Site A was very close to a major warren.

Table 4. The Cover and Height of the Vegetation in the Three Treatments (X, Y, and Z) in 1976 and 1982

Vegetation cover (%)	Site	1976	1982	Difference		Vegetation height (mean, cm ± 95% c.l.)	
				t_s	p	1976	1982
X-ungrazed	A	97	92	1.57	>0.1	17.1 ± 4.2	18.1 ± 2.4
	D	96	93	0.97	>0.2	16.5 ± 2.5	20.9 ± 1.5
Y-grazed rabbits	A	91	73	3.42	<0.001	10.7 ± 3.7	5.7 ± 1.2
	D	96	90	1.70	$0.1 > p > 0.05$	14.7 ± 3.3	16.6 ± 2.0
Z–grazed rabbits and horses	A	87	56	5.02	<0.001	8.2 ± 1.8	2.8 ± 0.8
	D	93	71	4.26	<0.001	15.1 ± 2.8	9.6 ± 0.9

Note: There was no change between years in the ungrazed, but a significant decline in both cover and height in the plots grazed by both species (Z). The rabbit-only treatment (Y) was similar to treatment Z in site A, intermediate in site D.

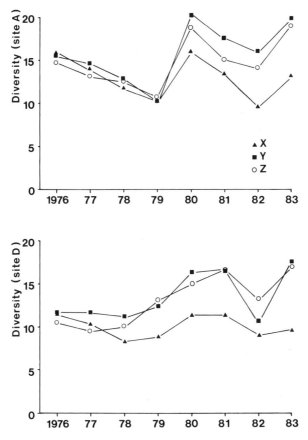

Fig. 10. The mean species diversities of the treatments in sites A and D. The ungrazed (X) sites did not change significantly between 1976 and 1983, but the grazed (Y, Z) ones increased in diversity (sign tests, $p = 0.06$ for Y exclosures, $p = 0.03$ for X exclosures; symbols as in Fig. 2 Graph a).

In addition to affecting the structure of the vegetation, grazing had marked effects on the species composition. The taller plants of the ungrazed treatment stood out in a sea of flowers of the annual *Bellis annua*; see photographic plates.

Plant species diversity fluctuated during the years 1976–1982, but there were no significant trends ($r_s < 0.86$, $N = 7$, $p > 0.05$; see Fig. 10). There was no clear effect of the treatments until 1980: Thereafter, the vegetation in the two grazed treatments became more diverse than in the ungrazed ones. The effect of rabbits was clearly the dominant one, for the Z treatment never differed significantly from the Y. Species richness responded differently, declining in the ungrazed and increasing slightly in the two grazed treatments, but the changes were not statistically significant.

Table 5. Contributions to the Herb Layer of Three Treatments

	Contribution (Median, %)			
			Z	
		Y	Grazed	Significance
	X	Grazed	(rabbits +	level
Site	Ungrazed	(rabbits)	horses)	(sign test)
A				
Principal food				
plants of the horses	77.0	43.8	31.8	<0.02
Annuals	1.2	1.7	21.2	<0.02
Limonium spp.	15.2	18.1	22.6	>0.05
D				
Principal food				
plants of the horses	79.8	63.3	42.2	<0.02
Annuals	6.2	4.4	18.6	<0.02
Limonium spp.	1.9	2.9	6.5	<0.02

Note: The three treatments were (1) principal food plants (*Agropyron, Carex, Dactylis, Bromus madritensis, Halimione*), (2) annual species, and (3) *Limonium* species (unpalatable perennials). Contributions measured as touches on point quadrats; significance levels are given for the differences between Y and Z treatments; data from 1981 and 1982 are combined.

The principal food plants of horses, which occurred in the study plots (see Chapter 4, Table 2) perennial or biennial grasses and forbs, were all commoner in the ungrazed and rabbit-grazed than in the horse-grazed treatments; see Table 5. Their place in the treatment grazed by horses and rabbits was occupied by annual grasses and forbs and by perennial halophytes, *Limonium* spp., which were not important food plants.

These differences between the ungrazed and grazed treatments arose in a wide variety of ways. Some principal food species, such as *Dactylis glomerata*, were rare at the start of the experiment and became very common in the protected plots, but not in the grazed; see Fig. 11. Others such as the biennial *Bromus madritensis*, common at the beginning became progressively rarer in the grazed plots. In the ungrazed plots, their frequency did not increase significantly.

For the annual plant species, almost all the species tended to decline in the protected plots, while still maintaining themselves, sometimes even increasing under grazing; see Fig. 12, *Bellis annua*. Other annuals, initially rare, became very common in the grazed plots, such as *Anagallis arvensis* (see Fig. 12).

With the projected increase in the numbers of horses, there was concern that overgrazing might lead to a decline in the productive capacity of the herbaceous plants. This was monitored in all facets except the deep marsh between 1979–1985, by clipping PC samples (see Appendix 4, p. 222) after

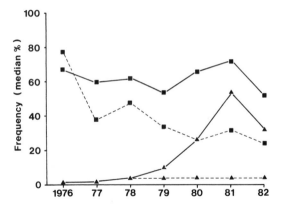

Fig. 11. Trends in the abundance of two important food plants of the horses in protected plots (solid lines) and under grazing by horses and rabbits (dotted lines): *Dactylis glomerata* (triangles) increased in frequency when protected from grazing ($r_s = 0.96$, $N = 7$, $p < 0.05$); in *Bromus madritensis* (squares), protection from grazing had no effect, but grazing caused the frequency to decline ($r_s = -0.93$, $N = 7$, $P < 0.05$).

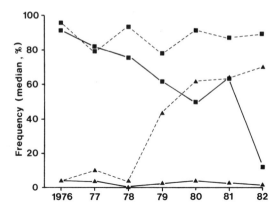

Fig. 12. Responses of two annual plant species to the treatments. Solid lines-ungrazed; dotted lines-grazed by horses and rabbits: in *Bellis annua* (squares), protection caused a sharp drop in frequency (see Plate 1, $r_s = 0.89$, $N = 7$, $p = 0.05$); but there was no change under grazing. In *Anagallis arvensis* (triangles), grazing caused an increase in frequency ($r_s = 0.94$, $N = 7$, $p < 0.05$), but there was no change in protected plots.

a whole growing season (January–December, see Appendix 4). The yield (Y) varied from an average of 800 g/m^2 in the fallow fields to 200 g/m^2 in halophyte grassland. Precipitation (P, mm) accounted for 60% of the variance between years

$$Y = 0.58\,P - 303, n = 7, p < 0.05,$$

and there was no correlation between the density of horses, or horses plus rabbits, in the previous year and the residual variance.

Production under grazing in 1982 was not significantly different from the peak standing crop in the annual exclosures (179 vs. 176 g/m^2; $t = 0.27$, $df = 7$, $p > 0.5$). I therefore conclude that the grazing pressure of horses and rabbits did not affect plant production (by depressing or stimulating it) in any straightforward way, even at the high densities observed in 1982 (see Chapter 1, Fig. 6 and Chapter 2, Table 5).

Changes in the Cover of *Phillyrea* scrub

The dominant shrub species on the grasslands was *Phillyrea angustifolia*, a species not eaten at all by the horses and eaten little by the rabbits (Chapter 4, Rogers 1979).

The cover of *Phillyrea* bushes in coarse and halophyte grasslands in 1974 was measured by J.-M. and M. C. d'Herbès as 6.2% (4.8–7.8, 95% c.l.). In 1980, it was 12.1% (10.2–14.3, 95% c.l.). There was no suggestion that the rate of increase was less in areas of high shrub density; the slope of the regression of cover in 1980 on cover in 1974 was greater than 1 (1.27 ± 0.21, 95% c.l.). Were the increase to continue at the same rate, the whole of the grasslands would be covered in scrub by the year 2000. Increases in browse production under grazing by horses has also been noted in rangelands in the United States, where the use of horses to favor the production of browse for deer is recommended as a management strategy (Reiner and Urness 1982).

Discussion of the Impact on Plants

Successional Changes When Grazing Was Removed

The range used by this herd had a long history of grazing by sheep and horses and of browsing by sheep and goats. From 1974, only horses used the area, and their preferred food plants were monocotyledons, especially the wetland species. Their impact, much stronger on the wetland vegetation than on the dryland, affected not only the structure and species composition, but also the aerial net primary production (ANPP).

In each site, the perennial monocotyledons increased when grazing was removed. The rate of change was fastest in the most productive habitats, the marshes: *Phragmites* dominated the deep marsh, and *Scirpus maritimus* the shallow marsh plots after 6 years.

In the grasslands, the changes were much slower. Neither species diversity nor richness changed significantly over 7 years, but perennial species did become more important. In the medium term, tall grasses such as *Dactylis* and *Agropyron* and the tall-growing forb *Halimione* will probably exclude the majority of other species.

The horses were the only grazers with a significant impact on the marshes, so it was certainly they which checked the development of *Phragmites* beds. In the grasslands, the eruption of the rabbit population in 1979–1980 meant that they and the horses removed similar amounts of herbage by 1982. The main foods of the rabbits in the growing seasons were annual forbs and grasses (Rogers 1979), so the horses were probably the main check on dryland as well as wetland perennials.

Because sheep and goats were removed from the pasture, *Phillyrea* has spread rapidly and, at present rates of increase, would cover the high ground by about the year 2000. In the long term, therefore, there is little doubt that this shrub will eliminate the grasslands unless a browsing herbivore is added to the system.

The kinds of change which followed the removal of grazing obey the same principles that govern changes in ungrazed vegetation anywhere and especially in the marshes of northern Europe (Ranwell 1972, Bakker 1985). The removal of the animals leads to a reduction in the disturbance to the plant community and fewer regeneration gaps (Crawley et al. 1987). Increasing height and cover reduce opportunities for germination; this leads to a rapid decline in the numbers of annual plant species. Heightened competition leads more slowly to further reductions in species richness (Harper 1969, Grubb 1977; Grime 1977). These changes initially favor perennial monocotyledons; in the very long term, these are in turn replaced by woody plants.

Without large mammalian herbivores, it is probable that the Camargue study area would eventually be covered by three main vegetation types, each more or less completely occupied by one or two perennial plant species: marshes (*Phragmites* and *Scirpus maritimus*), salt flats (*Arthrocnemum* spp.), and high ground (*Phillyrea angustifolia*).

An interesting difference between succession in the marshes of the Camargue and of northern Europe is the very slow colonization of the marshes by woody plants. In the north, *Phragmites* marshes are rapidly colonized by trees such as alder (*Alnus* spp.) and silver birch (*Betula pubescens*, e.g., Lecomte et al. 1981, Lecomte and LeNeveu 1986). In the Camargue, the only woody plant which colonizes the marshes to any extent is the tamarisk (*Tamarix gallica*); *Salix*, *Fraxinus*, *Populus*, and other wetland trees do occur but are restricted to the edges of rivers and freshwater canals. The factors which prevent the colonization of marshes by these species are not well understood; they are perhaps related to the slow rate of peat formation in the high temperatures of the Mediterranean zone, and to the presence of relatively high salt concentrations in the subsoil water of the middle and lower Camargue (Heurteaux 1970).

Changes in the Grazed Plant Communities Since 1974

The main changes occurred in the marshes. The increased numbers of horses caused the near elimination of *Phragmites*, and a decline in the height of *Scirpus* and in its density in the deeper areas. The species which increased were annuals, *Characeae* and *Ranunculus baudoti*. *Scirpus maritimus* maintained itself more successfully, especially in the shallower waters (e.g., Fig. 3), where its density declined only in 1982 under the very heavy grazing pressure of that year. In the deep marsh site, on the other hand, *S. maritimus* declined from 1979 to 1983, probably because of competition for light with the submerged vegetation, which formed a dense mat in the deep water early in the year.

The horses did not eat the submerged vegetation, so an important effect of their activities was to reduce the productivity of the marsh, and especially of their preferred emergent plants. A similar effect on production has been found with other wetland plants (Smith and Kadlec 1985).

In the grasslands, the pattern was the same, though the species involved were different. Height and cover declined under grazing, while diversity increased. Several of the perennial species declined in the plots grazed by horses, while annuals increased. The only perennials which increased were *Limonium* species, which were not important food of the horses. Interestingly, the introduction of horses into this range did not lead to the appearance of so-called latrine areas, a universal feature of horse grazing on pastures (Carson and Wood-Gush 1983, see also Putman 1986).

Why did these changes occur? This study provides no information on the mechanisms involved, but they are presumably the same as elsewhere. Success in competition for light, nutrients, and water in any plant community depends on rapid growth early in the growing season. In perennial species, this is promoted by reserves in the root systems. Grazing (or clipping) reduces the root systems of grasses (see Dittmer 1973) and also increases the probability of a plant dying in droughts (Box 1967), such as occurred in the Camargue in 1981–1982.

As predicted, the horses had a greater impact on the marshes than on the other habitats. Throughout the growing season, from the first appearance of the shoots, the horses used *Phragmites* very intensively: it was one of their preferred species throughout the warm season. The effects of grazers on a plant depend on the number of shoots it has, and on the frequency, extent, and timing of defoliation of the shoots. Grazing has its smallest effect on plants with high densities of shoots, which are lightly and infrequently grazed after the growing season. The density of reed shoots is much lower (tens per m²) than shoots in grassland swards (hundreds per m²). As a result, each shoot in a marsh is defoliated much more often than in a grassland and at a time of year when the plant is at its most sensitive. Furthermore, in most grasses, the meristem is at ground level, which protects it against all but the closest grazers; in *Phragmites*, the meristem is located at the internodes. If the shoot is bitten off below the water's surface, water enters the stem and may cause rotting of the meristem and other tissues. Comparable effects of grazing on this species have been found with cattle and horses in northern Europe (van Deursen and Drost 1990).

The horses' impact on the wetland vegetation was already noticeable when their food intake was only about 5% of the primary production (see Chapter 7, Fig. 2). By 1982, the horses were eating about 20% of production and rabbits perhaps half again (see Chapter 2, Table 5). The impact of these mammals was such that the abundance of perennial grass species in the deep marsh and in the grasslands declined sharply. As the horses preferred the large perennials, the effect of heavy grazing in the later years was to reduce the food value of the swards: these changes in the species composition of the plant communities were certainly the cause of the decline in the quality of the horses' diet between 1975 and 1983 (see Chapter 4, Fig. 7). Continued pressure of this kind would, in the long term, lead to the establishment of vegetation communities with a very low capacity to feed large mammals.

This kind of strong effect of grazing has been documented for horses in other wetland ecosystems (Eline and Keiper 1979, Turner 1987) and for guilds of large herbivores, wild as well as domestic (McNaughton 1984, Putman 1986, Milchunas et al. 1988). In the New Forest, free-ranging ungulates, including horses, have such powerful effects on vegetation structure and species composition that they prevent natural regeneration over large areas of the Forest.

Effects on Plant Production

In the New Forest, grazing (as practiced there) reduced the productivity of swards by c. 30% (Putman 1986, page 163). Moderate grazing can increase the productivity of plant communities (McNaughton 1984). Among the mechanisms involved, fertilization by feces has been shown to increase ANPP in wetland grasses (Hik and Jeffries 1990). However, wetland plants are very sensitive to defoliation and trampling (e.g., Turner 1987). ANPP is frequently reduced by defoliation and breakage due to trampling (e.g., Smith and Kadlec 1985, Kerbes et al. 1990, van Deursen and Drost 1990).

The effect of grazing on production of emergent wetland plants was dramatic—it caused them to produce only 20% of their potential. In this study, grazing had little effect on production in the grasslands, but continued heavy grazing pressure would have done so, by reducing the plant cover.

Effects on Invertebrates and on Other Mammals

Grazing leads to an increase in the abundance of soil invertebrates in some ecosystems (Granval et al. 1988) and to a decline in others (King and Hutchinson 1976). The very low densities in the New Forest may have been caused by the heavy grazing pressure of horses and cattle (Packham, in Putman 1986). A short-term study in the Camargue showed that the removal of grazing caused no change in the total biomass of soil meso-invertebrates, but equitability declined (Poinsot-Balaguer and Bigot 1980).

No data are available on the changes in numbers of invertebrates of the herb layer and soil surface, but it is probable that the heavy grazing pressure of the later years caused many species to decline. In temperate grass-lands, Hemiptera are more abundant and diverse when grazing or cutting ceases for some years (Morris and Plant 1983). An increase in the grazing pressure by sheep has also been shown to cause a decline of pasture insects (Dixon and Campbell 1978, Hutchinson and King 1980). A comparison of protected and very heavily grazed plots in the New Forest, on the other hand, found different communities, but about the same abundance of ground invertebrates (Putman et al. 1989). The invertebrates of open habi-tats such as the *Carabidae* were abundant under grazing, while those which depend on dense herb layers, such as some species of *Phoridae*, were rarer.

Among the rodents, the densities of *Apodemus sylvaticus* were similar in the study area and in an ungrazed area, but no *Microtus agrestis* at all occurred in the study area, whereas they were common—there may be 20 individuals/ha in some seasons—in the absence of horses (M. Jamon, per-sonal communication, Feb. 1982). *Pitymys duodecimcostatus* were found in the protected plots of the exclosures and nowhere else. It is probable that the horses, by reducing the cover and height of the vegetation and perhaps by compacting the soil, rendered the grazed area unsuitable for these small mammals. *Apodemus* are insectivorous and granivorous and live in *Phil-lyrea* shrubs, whereas the voles (*Microtus*) are grazers which prefer dense grasslands.

Similar conclusions were drawn by Putman et al. (1989) from a compari-son of protected and grazed plots in the New Forest: *Clethrionomys* was common in the ungrazed but never trapped in the grazed treatment; and *Apodemus* occurred in both, though they were much more abundant in the ungrazed.

Wild boar, too, showed a notable interest in the exclosures and ploughed them to eat the bulbs of the protected *Scirpus maritimus*. There is little doubt that the root systems of these ungrazed plants were denser than those of the sparser, grazed ones outside. Whether the horses had any effect on numbers of wild boar is matter for conjecture.

The commonest mammal on the pasture, the rabbit, increased in num-bers as the horses increased (see Chapter 2, Table 5), but it is unlikely that this was caused by facilitation (Oosterveld 1983) because the increase occurred both in the study area and where there were no horses (Vande-walle, personal communication, May 1988). Rabbits do not require a dense herb layer for cover or for feeding. A rabbit weighs less than ½% the weight of a horse and can therefore tolerate a much lower rate of food intake per individual. When food is limiting, the smaller species should outcompete larger (Clutton-Brock and Harvey 1983). There seems little doubt that the offtake by the rabbits (see Chapter 2, Table 5) contributed to the decline in performance by the horses in 1982 (see Chapter 1, Fig. 6).

In conclusion, heavy grazing by large herbivores such as the horses naturally leads to a decline in the abundance of animals which require cover (some terrestrial invertebrates and rodents). Species which prefer open habitats, including small herbivores such as the rabbit, are facilitated by horses under some circumstances.

Effects on Birds

Counts of the breeding birds in the study area and the ungrazed area next to it, by Olivier amd Jean-Pierre Biber (personal communication, June 1986), showed that the increase of the shrub layer was accompanied by a very strong increase in the breeding passerine population. The only species which showed an increase which could be imputed to grazing by horses was the yellow wagtail. Very few birds depended on the marshes for breeding; of these, some declined, such as three species of reed-nesting warblers (five pairs) and the single pair of marsh harriers. Coot continued to nest, but in tamarisk bushes on the edges of the marshes rather than in emergent vegetation (Salathe 1985). In other ecosystems, increases in breeding populations of passerines have also been reported after increases in the shrub and herb layers caused by the relaxation of grazing (Taylor 1986).

Other breeding birds which used the marshes for feeding were the herons, *Ardeidae*. Their feeding habitats differ among species, little egrets and grey herons feeding in open waters, while the others, purple, squacco, and night herons, feed in more closed habitats (Bauer and Glutz 1966). The progressive opening up of the marsh would therefore be expected to favor the little egret and grey herons rather than the others. No data are available on use by these species, but fewer purple herons were seen in the later years, while for little egrets, the marsh became one of the principal feeding areas on Tour du Valat during the breeding season (H. Hafner, personal communication, June, 1985). Even in winter, considerable numbers of little egrets and grey herons used the marsh, see Fig. 13.

Cattle egrets use livestock as beaters for their prey (Siegfried 1978) during the seasons when they feed on invertebrates and amphibia. Large numbers foraged around the horses in the warm seasons; see photographic plates.

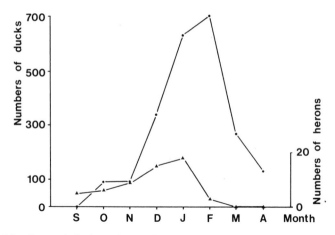

Fig. 13. Numbers of duck and coot (circles); and herons and egrets (triangles) wintering on the deep marsh, 1983–1984.

The use of the marshes in winter by coot and by dabbling ducks (teal, gadwall, shoveler, and mallard) was initially trivial and increased over the years until several hundreds of birds regularly used the open part of the deep marsh (20 ha, Fig. 13). These species feed on a wide variety of foods (seeds, oogonia of *Characeae*, zooplankton, and the vegetative parts of submerged plants, Tamisier 1971, Pirot et al. 1984, Allouche and Tamisier 1984). They all feed in open water, so the progressive opening up of the emergent vegetation improved the structure of the habitat for them. The same principles therefore govern the impact of heavy grazing on birds, as were described previously for the invertebrates and rodents. In addition, the great increase of *Characeae* would have further favored the species which feed on these plants, teal and gadwall.

Horses and the Management of Wetlands in the Camargue

The results reported here show that extensive grazing by horses, even with no supplementation at all, had a profound effect on the nature of the wetland plant communities, on their productivity, and on the way they were used by waterfowl.

Whyte and Cain (1981) warned that uncontrolled grazing was usually detrimental to breeding waterfowl populations because the cattle remove too much of the cover and food plants of the waterfowl. Rotational grazing allows much higher waterfowl production than season-long grazing (Gjersing 1975); and many authors have argued for the adoption of rotational systems in areas used for both breeding waterfowl and cattle production (Whyte and Cain 1981, Holechek et al. 1982b).

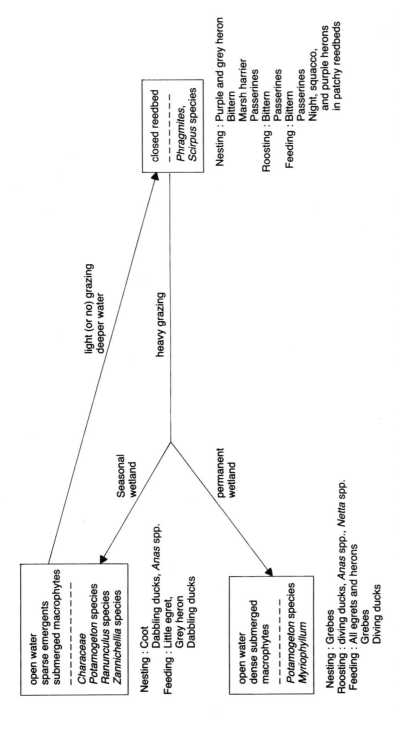

open water	**closed reedbed**
sparse emergents	
submerged macrophytes	*Phragmites,*
	Scirpus species
Characeae	
Potamogeton species	Nesting : Purple and grey heron
Ranunculus species	Bittern
Zannichellia species	Marsh harrier
	Passerines
Nesting : Coot	Roosting : Bittern
Dabbling ducks, *Anas* spp.	Passerines
Feeding : Little egret,	Feeding : Bittern
Grey heron	Passerines
Dabbling ducks	Night, squacco,
	and purple herons
	in patchy reedbeds

light (or no) grazing
deeper water

heavy grazing

Seasonal
wetland

permanent
wetland

open water
dense submerged
macrophytes

Potamogeton species
Myriophyllum

Nesting : Grebes
Roosting : diving ducks, *Anas* spp., *Netta* spp.
Feeding : All egrets and herons
Grebes
Diving ducks

Fig. 14. Schematic representation of the role of grazing by horses and cattle in the determination of the community structure of birds in the freshwater marshes of the Camargue.

The bird communities of freshwater and brackish wetlands of the Camargue are complex and include both wintering and breeding populations of herons, ducks, coot, birds of prey, and passerines. Relationships between the plant and bird communities, and the role of grazing mammals in maintaining diverse early successional stages are schematized in Figure 14. In addition to the three illustrated, there is a wide range of intermediate types; it is in these that purple, night, and squacco herons feed. Species which use a range of dryland and wetland habitats have been omitted (e.g., black-headed gulls, pratincoles, and gull-billed terns).

The species favored by heavy grazing in the later years of this study are the commonest ones in the Camargue as a whole—the little egret among the herons and teal among the ducks (Tamisier 1972, Hafner 1977). The marshes of the Camargue have been grazed by horses and cattle since time immemorial and still are, wherever the salinity is low enough for emergent plants to grow. The large mammals are an overridingly important determinant of the present structure of the waterfowl communities, through their impact on the vegetation.

Conclusions: The Impact of Equids on Plant and Animal Communities

The difficulty of separating the effects of equids from the effects of other species of large mammal herbivore in guilds of grazers in natural ecosystems means that there is very little information available on wild equids. However, the fact that the digestive processes and diets of domestic and wild animals are so similar makes it likely that the same principles apply to all species of the family.

Horses, being graminoid specialists with a special liking for *Phragmites*, have a very strong effect on graminoids, and on wetlands. This results in both a reduction of the standing crop of the herbaceous plants and the creation and maintenance of open habitats especially in the marshes. Apart from trees, which they can kill by ring-barking, horses have little impact on woody plants, and these may invade grasslands heavily grazed by horses.

The consequences of these effects of horses on wetlands and on grasslands for other animal species are considerable: they provide managers of grazing lands with a powerful tool which is natural and productive, compared with many of the alternatives—herbicides, mechanical cutting, and fire, to name only three (Duncan and d'Herbès 1982, Gordon et al. 1990). The strong impact of equids on plants also means that feral populations that are not closely managed are considered a scourge, especially in North America (NRC 1982), by other users of the grazing lands with objectives which do not coincide with those of the equids.

Chapter 9 Equids in Grazing Systems

Une jument mange pour trois vaches.[a]

Saying of Camargue *gardians* (cattle herders).

In this book, I have described the interactions of a population of equids with their food resources. In most circumstances, equids coexist with other ungulates in grazing systems. Their interactions with the other species, especially with the ecologically close grazing bovids, are key components of the functioning of the grazing systems. The considerable overlap between equid and bovid diets (see Chapter 4, Fig. 11) will lead to competition if either of the populations is food limited.

In this chapter, I review the information available on the factors limiting equid populations, and on the nature of interactions between sympatric populations of equids and grazing bovids. I then draw some conclusions from this and other studies for the management of equid populations.

Factors Limiting Equid Population Sizes

Information on the mechanisms of population limitation is available for two feral horse populations and two populations of zebras. The largest remaining zebra population, the 200,000 individuals in the Serengeti eco-

---

[a] A mare eats for three cows.

system, has remained stable since 1960 and is clearly uncoupled from its food supply (Sinclair and Norton-Griffiths 1982). Predation on this population (e.g. by lions, hyenas, wild dog etc.) is much higher than on the food-limited ruminants and is sufficiently heavy—>60% of the annual adult mortality—to prevent the population from increasing.

The zebra of the Kruger National Park also live in an ecosystem with large numbers of predators. The population declined during a drought in 1969 (coupled with cropping) from 13,000 to 7,500 in 1975. In the latter part of this period, rainfall had returned to normal, and predation was presumed to be preventing the zebra from increasing. The considerable body of circumstantial evidence was backed up by an experimental reduction in predator numbers (lion) in one sector, which led to an increase in the group sizes of the zebras (Smuts 1976a).

Though the effect of predator control on population size was not reported, predation is clearly an important mortality factor in both these populations. The food supply cannot, however, be completely excluded as a regulating factor until further data are available, e.g., on seasonal cycles in body condition. Predation may be only the proximate cause of mortality.

The mountain zebra population of Namibia, on the other hand, appeared to be resource limited in the 1960s: recruitment rates of young animals were clearly low in dry years, and "during the drought of the 1968/9 season hundreds of Hartmann zebra were found dead or dying in the Khomas Hochland. . . . Normally (the zebra) are very wild and shy and will make off whenever they spot a vehicle. During this abovementioned drought however, they wandered aimlessly around. . . . One could drive to within fifty yards of them" (Joubert 1974, p. 51).

Many feral domestic populations live in areas where there are few or no predators of large mammals are increasing rapidly (e.g., in the Granite Range, Berger 1986) and are not yet subject to limiting factors. However, one of these population, the Sable Island horses (Welsh 1975), is naturally regulated at around 200 individuals. Each winter, the animals' food supply was so poor that they were in energy deficit, and in some years, harsh weather increased the number dying of starvation to the extent of causing a population crash. Without management, the Camargue population, too, would have been limited by mortality induced by starvation.

Long-term quantitative studies of the factors regulating free-ranging equid populations, which allow firm conclusions to be drawn are urgently needed. However the preceding information presented above suggests that populations of equids of mesic habitats can be either predator or food limited. Predator limitation may be the rule in natural ecosystems with abundant large mammalian predators, and food limitation the rule in artificial ecosystems. There is no evidence for an intrinsic ("social") mechanism that regulates native (wild) or feral equid populations below the levels determined by their resources.

Interactions with Other Grazing Ungulates

In Natural Grazing Systems

Early descriptive studies of communities of African ungulates showed that there is a high degree of ecological separation among sympatric species in the habitats they use (Lamprey 1963) and in their diets (e.g., Field 1972). The principal distinction was between browsers and grazers. Within grazers, there was overlap in resource use, but still ecological separation (Stewart and Stewart 1970). It was widely accepted that the evolutionary mechanism behind this separation was interspecific competition (e.g. Sinclair 1979, see also Owen-Smith 1985).

Bell (1969, 1970) drew attention to the dynamics of resource use by the grazing species. They overlap in their use of habitats, and plants, but these animals have a predictable grazing succession, with the large species (elephant and buffalo) being the first to use the tall grasses in the lower parts of the catena. These large species and the zebras eat more of grass stem than do the smaller species, such as wildebeest and Thomson's gazelles which follow them. The smaller grazers eat more of the higher quality leaves. Bell's hypothesis that the smaller species depend on the larger ones to open up the tall grasses revolutionized concepts of the ecological relations between the equid and the grazing bovids: the equids were seen not as competitors, but as an integral part of the bovid-dominated grazing system which allows the smaller bovids to exploit coarse swards. The diets of zebra and the much larger buffalo are closely similar (Gwynne and Bell 1968, Sinclair and Gwynne 1972).

Recent data from Bell's study area, the Serengeti ecosystem, has allowed these concepts to be tested. Sinclair and Norton-Griffiths (1982) show that the wildebeest have increased by a factor of five since 1960, to about 1 million. The zebra have remained stable at c. 200,000. Facilitation by zebra is clearly not necessary for wildebeest: further, competition for food is not likely to be an important process in the regulation of zebra numbers.

Wildebeest may not be competing with the zebras, but are the zebras competing with the food limited wildebeest (Sinclair et al. 1985)? They do not use exactly the same habitats—zebras spend more time in the tall grasslands and in riverine habitats—but they overlap (grass heights × habitats, Sinclair et al. 1985, Fig. 3) by

$$0.80 \times 0.55 = 0.44.$$

Thus $200,000 \times 0.44 = 88,000$ zebras occupy wildebeest habitat. These animals could eat $170g/W^{0.75}/day$ (see Chapter 3). If the population mean body weight of zebra is 200 kg (Sinclair 1975), the grass consumed would be 9 kg/individual/day; 88,000 zebra would therefore consume 792 tons/day in wildebeest habitat. The average wildebeest requires 4.25 kg/day (Sinclair 1975). The zebras can therefore eat food which would maintain

$$792,000/4.25 = 186,000 \text{ wildebeest}$$

or some 15% of the current population.

In this, the best-studied near-natural ecosystem, the equids' relations with the dominant species of sympatric grazing bovids are not facilitative; they are competitive, but only in one direction, because the equid population is apparently not regulated by its food supply.

As shown in Chapter 3, the persistence of equids since the radiation of the evolutionarily younger bovids is likely to be a consequence of their ability to extract nutrients from forages at a higher daily rate. In Chapter 3. "General Discussion" I outlined three mechanisms by which the medium sized grazing bovids could nonetheless dominate grazing systems.

Inadequate food supplies on the range could mean that equids are not able to realise the high daily rates of nutrient extraction of which they are capable in circumstances where bovids do. The data from the Serengeti do not support this hypothesis, because there the food supply is limiting for the grazing bovid, but not for the equid. Nonetheless this mechanism may be important on other ecosystems: further field data are necessary to test this hypothesis.

Bovids may be able to use a wider range of plant tissues, particularly dicotyledons, than equids. This is the case in at least some ecosystems (Chapter 4, "General Conclusion"). However the consequences for their nutrition and population processes remain to be measured.

Finally, the success of the medium sized grazing bovids may be due to their greater ability to evade predators. The relatively short feeding time of bovids (Chapter 5, "On Equid Foraging Times") means that they can feed and then lie-up in places favorable for the surveillance of predators while the equids are still feeding, and presumable more exposed to predation. This hypothesis is consistent with the data from the Serengeti (see above "Factors Limiting Equid Population Sizes"), where predation rates on zebra are sufficient to limit zebra numbers, but not wildebeest. How common this situation is will be known only when further field studies provide new quantitative information on the regulation of populations of sympatric equids and bovids living in near-natural ecosystems.

The principal interaction between equids and grazing bovids is therefore competition. There is evidence that all the three mechanisms proposed in Chapter 3 contribute to the evolutionary success of the bovids, and it is likely that their interactions will differ among ecosystems, with the impact of the large mammalian predators being a key determinant of their nature. Further field studies are urgently needed to provide data to test current theory.

In Artificial Grazing Systems

The most detailed study has been conducted in the New Forest, in the United Kingdom, where the animals are managed extensively. Resource use by cattle overlaps with the horses by 0.55–0.77 (Pianka's index, habitats and diets combined; Putman 1986, Table 4.8), according to the season. Though some animals are removed by their owners, their poor body condition implies that their numbers are limited by food availability. The high overlap indicates that they are experiencing competition (MacArthur and Levins 1967).

As the quotation at the head of the chapter indicates, pastoralists in the Camargue perceive horses as competitors of cattle. A mare can eat

$$0.170g \times 425kg\ 0.75 = 16kg/day$$

while a Camargue cow eating 2% of her liveweight requires

$$0.02 \times 275kg = 5.5kg/day.$$

If their resource use is identical, then a mare indeed eats food which would maintain $16/5.5 = 2.9$ cows, or 2.2 cows if their dietary overlap is 0.75 (cf. Chapter 4, Fig. 11).

When food is limiting, as is generally the case in pastoral ecosystems, then equids will compete with cattle, and they may have a very strong impact on the plants, on species composition, cover, and productivity.

In the Serengeti, the vast ungulate populations do not cause obvious overgrazing (reduced cover and/or productivity), apparently because some of the species are maintained below the food ceiling by predation; also, most important, the commonest species have a seasonal migration, which means that the major part of the ecosystem's grasslands are grazed lightly during the growing season. This allows the plants to reproduce and reconstitute the resources in their roots before being grazed heavily in the dry and early wet season.

Feral equids can reach much higher densities relative to their resources: they have no large mammal predators and their home ranges are much smaller (<10–$50km^2$, Berger 1977, 1986, Feist and McCullough 1976, Rubenstein 1986) than those of zebras in near-natural ecosystems, ($100 \geq 1000kg^2$, Klingel 1967, Joubert 1972, Smuts 1975a, Maddock 1979). Compressed by human activities, they do not often show seasonal migrations (but see Berger 1986) and therefore exert their grazing pressure on the forage plants in the growing season. This can lead to overgrazing (i.e., reduced plant cover in arid areas, Carothers 1976, Hanley and Brady 1977; and reduced plant productivity, e.g., Putman 1986 and the Camargue study).

Interactions between equids and grazing bovids in artificial as well as natural ecosystems are therefore principally competitive. The potentially high rate of food intake by the equids means that when food is limiting, equids can have a strongly depressive effect on the sizes of populations of sympatric bovids. This often leads to pressure to manage the equids.

Management of Free-Ranging Equids

Zebras are harvested for meat and skins in much of Africa. The only large population for which management (i.e., actions aimed at obtaining a defined change in population size) has been documented, is the Kruger National Park population. This increased, with wildebeest, during the dry 1960s. The park authorities, fearing overgrazing, culled both grazers; they declined by c. 50%. "Despite a phenomenal improvement in grass cover, the rapid relaxation of cropping procedures and their eventual termination in 1974 the declines persisted" (Smuts 1976a, p. 99).

The failure of these grazers to increase may have been due to a change in the quality of the grass swards; and in particular to a reduction in the area covered by short grasslands which heavy grazing had contributed to maintaining. Management of wild populations in grazing systems requires a more detailed understanding of the functioning of most ecosystems than is currently available. Further descriptive and analytical studies of equids in near-natural ecosystems are needed: a priority is surely to test the hypothesis that the Serengeti and the Kruger National Park zebras are regulated by predation, below the level which could be maintained by the food supplies.

Management of free-ranging equids has focused on three problems: first, the conservation of the declining populations of wild species such as Asian asses and mountain zebras; second, reduction of the numbers of overabundant feral horses and asses; and finally, in parts of the world where wild ungulates have been eliminated by humans, on the use of carefully managed populations of domestic or wild equids to manage the vegetation of nature reserves.

Conservation of Wild Equids

Five of the seven *Equus* species are threatened with extinction (the African wild ass, Grevy's and mountain zebra, the onager, and Przewalski's horse, IUCN 1988). Their conservation requires management to increase the remaining populations and to reintroduce them into parts of their ancestral ranges where conditions now allow them to maintain themselves. This work is coordinated by the IUCN Equid Specialist Group.

The limiting factors are not known with certainty for any of the threatened wild populations, but it is widely held that the main one is usually resource competition with domestic stock (e.g., Groves 1974,

Wolfe 1979, IUCN 1988), leading to high mortality rates. In view of the information presented above the preceding section, "*In Artificial Grazing Systems*," this is likely to be true.

All the threatened species would benefit from the establishment of new, secure, free-living populations, but reintroduction is needed most urgently for the Przewalski horse (Seal et al. 1990): there have been no confirmed sightings in the wild since the 1960s. The species has been saved from extinction by the establishment of a zoo population, which, in 1990, numbered c. 1000. These are descended from 13 horses, the last of which was caught in the wild in 1947.

Life in the zoo environment, without natural selection, will ultimately result in undesirable changes in the genome of the species. There is therefore an international effort to locate sites where free-ranging populations of this species can be established in conditions as close to wild as possible. The reintroduction program should be modeled on the approach proposed by Stanley Price (1989): the founders will need to be carefully chosen so as to be able to produce offspring as outbred as possible, in order to avoid the inbreeding depression (lowered breeding and survival rates), which is measurable even in zoos when inbreeding coefficients exceed 0.15 (Bouman and Bos 1979).

Because the number of founders will generally be limited, the females should be chosen such that they (a) have reached adult weight, and thus are able to produce high-quality offspring (see Chapter 7); and yet (b) have the longest life expectancy: thus, they should be c. 4 years old. Mortality should be minimized by stringent disease control at the time of introduction, and by minimizing the probabilities of disease and predation thereafter. Particular attention must be paid to predation by wolves, which are common in parts of the ancestral range.

The terrain should have swards dominated by monocotyledons. Wetland geophytes (reeds, sedges) provide good grazing in the warm season, but they need to be complemented by perennial grasses in the cool season (Chapter 4). Unless these plants grow all through the year, and thus maintain good quality, the area should be large enough to provide large quantities of forage so that the horses are able to practice their high-intake strategy, which demands c. 2½ tons of forage for a 400-kg horse over a 5-month period (from data in Chapter 3). On a rangeland with an average grass biomass of 100 g/m^2, half a square kilometer is needed to maintain a horse over 5 months of winter, allowing for 50% losses to other herbivores and decomposition. In much of the grazing lands of the Camargue, which are relatively productive, 1/10 km^2 is required.

Specialists are understandably concerned about the ability of zoo-bred animals to re-create a functional social organization once they are returned to the wild. The strategy used will, inevitably, be determined by what is possible in each project, but on the basis of this and other studies, a sensible approach would be to preform family units of 2–4 mares with an unrelated stallion. These should be kept apart from other horses for at least a month before release. An alternative could be to release first a group of nonbreeding females. When they have learned to use the terrain, immature males (<3 years) could be introduced. They are unlikely to cause damage when competing for the females, as they fight less intensely than adult stallions and in any case are often dominated by the females (Chapter 6).

Nonetheless in the Camargue study, an apparently natural social organization was re-created in the worst possible circumstances where a single adult male with all the breeding females was challenged by three 4- to 5-year-old bachelors in an enclosed area. Mortalities were few and concerned newborn foals, which are unfortunately very fragile. A similar experiment has recently been conducted with Przewalski horses. I had the privilege to see the results with its initiator, Dr. V. Klimov at the Askania Nova estate, USSR in 1990. Here, an adult bachelor was introduced into a 1500-ha paddock where a single stallion held 27 mares. Two herds (10, 17 females) resulted, and since then, another breeding herd (1 ♂, 12 ♀♀) and a bachelor herd (12 young ♂) have been added. As befits a mesic-habitat equid, there is no territorial defense, and the groups are often close together.

Whatever approach is chosen, great care will need to be taken to avoid the accidental deaths, infant killing, and induced abortion that can occur in unstable equid societies.

Overabundant Populations of Feral Equids

In many parts of the world with low human densities, horses and asses, either introduced or escapees, have succeeded in building up viable populations. The best known ones are in the United States, but there are examples on all the continents.

These feral equids are sometimes welcome as part of the cultural heritage, or because they are considered to fill a vacant niche (Martin 1970), but when they compete for resources with domestic herds or wildlife, then there is always pressure to reduce or eliminate the populations of feral equids (NRC 1982).

Management to reduce the size of a population can involve increasing mortality rates, reducing recruitment, or both. The approach adopted should be based on an evaluation of the importance of these two processes in the regulation of the population (Wolfe 1982, see also Caughley et al. 1987). It is clearly inefficient to eliminate large numbers of horses if the survival rates of the remainder increase in a compensatory manner.

Considerable efforts have gone into finding humane ways of reducing feral equid populations. Sterilization of males by vasectomy is technically feasible but costly and risky: a few males which slip through the net can be enough to cause the ultimate failure of the action (McCort 1979).

Sterilization of females (e.g., by ovariectomy) is certain to have an effect on recruitment; the main disadvantage of this approach is the enormous cost of capturing and ovariectomizing a feral mare.

Mortality rates can be increased indirectly by making limited resources inaccessible (water or high-quality food). Horses are easy to fence in or out, as they are reluctant to jump, and few crawl under fences like cattle do. A three-strand barbed wire fence 1.2 m high with posts each meter will stop a Camargue horse in virtually all circumstances.

Capture for sale is widely used for managing feral horse populations; the success of this method naturally depends on the costs and the perennity of the markets. The most widely used method for managing feral equids is the bullet. Killing is obviously the last resort for moral reasons; further, when the population is not eliminated, it may be necessary to repeat costly kills on an annual basis, and some populations even increase when harvested (Caughley 1977, page 200).

If killing is necessary, then it should be done humanely, avoiding wounding, and eliminating whole family units so as to minimize disruption of the social system, which can cause further distress and high rates of neonatal mortality (cf. Chapter 7, *Mortality of young horses*). Culling programs which lead to male-biased sex ratios (e.g., 68% male, Ganskopp and Vavra 1986) are also likely to have these effects.

Equids in Nature Reserve and Wildlife Management

In many parts of the world, managers of nature reserves are faced with problems of controlling successional processes in plant communities, which would otherwise lead to losses either of diversity or of target species populations in plant and animal communities. A well-known example is chalk grassland, where perennial grasses tend to outcompete rarer, prostrate species, and even the grasses are ultimately replaced by hawthorn scrub (Smith 1980).

These problems typically occur in ecosystems where the large wild ungulates have been exterminated by humans (e.g., in Europe), though scrub invasion is also reported from African national parks (e.g., Kruger National Park, Smuts 1976a).

In nature reserves which are small and easily managed, the vegetation has often been controlled by the reserve authorities using a variety of techniques, such as fire, cutting, and even herbicides, as well as the more natural approach of using grazing by closely managed herds of domestic animals (Gordon et al. 1990).

For management using grazing animals, the feeding preferences of equids mean that they are very useful for the control of monocotyledons such as perennial grasses in chalk grasslands and *Phragmites* reeds in wetlands. In mountain pastures, horses have been found to be effective in converting scrub to grassland (Jarrige 1980). They also control some species of trees, even ring-barking large individuals of willow and poplar.

Their effect on forbs and shrubs is, however, weaker than that of cattle (e.g., *Phillyrea* in the Camargue), and horses are recommended as management tools to encourage the dicotyledonous food plants of browsing wildlife species, such as deer in North America (e.g., Reiner and Urness 1982, Holechek et al. 1982b).

Horses have a number of advantages over cattle for this kind of management. They can be contained by much lighter, and therefore less costly, fencing; they are less aggressive and therefore easier to round up; and in many countries, they are not subject to obligatory prophylaxis and/or tests for contagious diseases, such as foot-and-mouth disease or brucellosis in cattle.

Domestic horses are now widely used to manage the vegetation of nature reserves in Europe (Gordon and Duncan 1988, Gordon et al. 1990, Van Wieren 1990). In the future, this could provide opportunities to use wild equids, and thus to maintain additional populations of endangered equids in more natural conditions than zoos while solving a problem of nature reserve management.

Appendixes

Appendix 1 Latin Names and Common Names of Extant Mammal and Bird Species to which Reference Is Made in the Text

(From Walker 1968, Groves 1974 and Peterson et al. 1967)

Mammals

Order Primata
hamadryas baboon	*Papio hamadryas*
gelada baboon	*Theropithecus gelada*
indri	*Indri indri*

Order Lagomorpha
rabbit	*Oryctolagus cuniculus*

Order Rodentia
wood mouse	*Apodemus sylvaticus*
coypu	*Myocastor coypus*

Order Carnivora
wolf	*Canis lupus*
lion	*Panthera leo*
spotted hyena	*Crocuta crocuta*

Order Proboscidea
African elephant	*Loxodonta africana*
Asian elephant	*Elephas maximus*

Order Perissodactyla
American tapir	*Tapirus terrestris*
Malayan tapir	*Tapirus indicus*
white rhino	*Ceratotherium simum*

Indian rhino	*Rhinoceros unicornis*
black rhino	*Diceros bicornis*
Przewalski's horse	*Equus przewalskii*
plains zebra	*Equus burchelli*
mountain zebra	*Equus zebra*
Grevy's zebra	*Equus grevyi*
onager	*Equus hemionus*
kiang	*Equus kiang*
African wild ass	*Equus africanus*
tarpan	*Equus ferus sylvestris*
Order Artiodactyla	
hippopotamus	*Hippopotamus amphibius*
wild boar	*Sus scrofa*
red deer	*Cervus elaphus*
bighorn sheep	*Ovis canadensis*
wildebeest	*Connochaetes taurinus*
impala	*Aepyceros melampus*
waterbuck	*Kobus defassa*
blackbuck	*Antilope cervicapra*
klipspringer	*Oreotragus oreotragus*
gemsbok	*Oryx gazella*
African forest buffalo	*Syncerus caffer nanus*
African savanna buffalo	*Syncerus caffer caffer*
Asian water buffalo	*Bubalus bubalis*
American bison	*Bison bison*
European bison	*Bison bonasus*
oryx	*Oryx beisa*
topi	*Damaliscus lunatus*
Thomson's gazelle	*Gazella granti*

Birds

bittern	*Botaurus stellaris*
night heron	*Nycticorax nycticorax*
squacco heron	*Ardeola ralloides*
cattle egret	*Bubulcus ibis*
little egret	*Egretta garzetta*
grey heron	*Ardea cinerea*
purple heron	*Ardea purpurea*
mallard	*Anas platyrhynchos*
teal	*Anas crecca*
gadwall	*Anas strepera*
wigeon	*Anas penelope*
shoveler	*Anas clypeata*
marsh harier	*Circus aeruginosus*
coot	*Fulica atra*

black-headed gull   *Larus ridibundus*
gull-billed tern   *Gelochelidon nilotica*
yellow wagtail   *Motacilla flava*
reed nesting warblers   *Acrocephalus spp.*
pratincole   *Glareola nordmanni*

Appendix 2 The History and Management of Camargue Horses

The Camargue is a small riding horse; see Table 1 and photographic plates. The horses are born colored but rapidly turn white with age, so that when they reach between 3 and 8 years, they are all white haired and black skinned. This is a consequence of the loss, with age, of pigmentation in the hair; it is controlled genetically by a double dominant at the Grey locus (see Evans et al. 1977); and is unusual in horses. The only other well-known case was described by Herodotus two millennia ago—the "all white" feral or wild horses of the Pripet (Polesye) Marshes in central Europe—and they also lived in wetlands.

The morphology of the skeleton of Camargue horses has excited a great deal of interest. Early students (Toussaint 1874), found that there are strong similarities between the skeletons of Camargue horses and those of wild horses which lived in central France in the Quaternary period. These include *Equus ferus solutreensis* of which hundreds of skeletons, dating from c. 18,000 B.C., have been found near Solutre north of Lyon, and *E. ferus gallicus*, which lived in southern France some 10,000 years ago.

Camargue horses are hardy and can live and breed on poor forage. They become docile when handled young, but some, lacking application and aptitude are difficult to train. Their gaits are often rough and uncomfortable, which is not surprising in view of their conformation.

Table 1. The Official Description of the Breed (Arrêté Relatif à la Race du Cheval Camargue).

The Camargue Horse: Its Characteristics

The Camargue horse is a hardy saddle horse.

The head is large, usually square, and well-set.

The brow is flat, the head is straight and the nose often receding and unobtrusive, which gives the impression of a convex profile.

The ears are short and well-separated with a wide base.

The eye is flush with the head, because the edges of the sockets are not prominent.

The cheek muscles are powerful.

The mane is often thick.

Adult, the coat is white, sometimes speckled.

The chest is deep, the shoulder straight and short.

The neck is short.

The limbs are strong and well-built.

The hooves are broad and the animal, sure-footed.

The knees are wide; the horse is well-jointed.

The back is short.

The thighs are large and well-muscled.

The rump is short, full and lightly sloping.

The tail, which is well attached, is carried low. The hairs of the upper part are very dense.

The height varies from about 1.35 m to 1.45 m (13½–14½ hands).

The weight varies between 300 and 400 kg.[a]

The Camargue horse may not reach adult weight until 5 to 7 years old.

It saves its energy for action—at rest Camargue horses may appear unspirited. Sensible, lively, agile, brave, and with great stamina the Camargue horse can withstand fasts as well as harsh weather and can carry out feats of endurance.

Source: Ministère de l'Agriculture, 17 March 1978.
[a] An underestimate.

Origins

Although the early fossil record for the equids is well represented, there is little or no material available for the crucial period around 5000 B.P., when domestication of horses took place. It is not known exactly when, where, and how this relationship between people and horses first developed. There is some indirect evidence, behavioral and morphological (Ebhart 1954, Groves 1974), that three or four different species or subspecies of horses were domesticated in parallel and then crossed to produce the wide range of breeds which exists today. Among these distinct types were certainly the ancestors of the Arabians and of the tarpans (*E. ferus sylvestris*), and there was probably a small tundra pony similar to the northern ponies of today and a large tundra horse rather like present day Andalusians.

This uncertainty about the origins of modern horses in general has not discouraged several authors from being dogmatic about the origins of the Camargue breed; the several hypotheses have been well discussed by Bérriot (1969). They boil down to two main ones, which propose on the one hand a prehistoric origin and on the other an escapee origin.

The first historical record referring to horses in the Camargue region is Phoenician (c. 2500 B.P.), and it mentions wild horses living on the sparse pastures on the edge of the Rhône and records that the horses were hunted for meat by the inhabitants of the delta (see Vlassis 1978, Allier 1980). These horses may have been wild descendants of *E. f. gallicus*, but the possibility that they were already escapee domestic animals cannot be excluded. Horses had been domesticated some 1500 years before (2000 B.C. in southwestern Asia), and there is no evidence for the presence of any wild populations in southern Europe after 5000 B.C.

The prehistoric hypothesis (Toussaint 1874), which holds that the horses found in the Camargue in the 19th century were a relict population of the Magdalenian *E. f. gallicus* is clearly false in its extreme forms because there were considerable movements of horses in and out of the region of Arles in Phoenician, Roman, and Saracen times (Vlassis 1978). Also, it is known that attempts have been made since the 17th century to "improve" the breed with imported stallions (Allier 1980), and there is no reason to suppose that this has not been happening since Roman times. Camargue horses of today are therefore principally or totally of domestic origin.

History of Camargue Horses: 16th Century to 1948

The period falls naturally into three sections, which are characterized by the different uses to which the horses were put and by differences in the practices of cross-breeding, selective breeding, and husbandry. The first period dates from the first modern historical references (de Quiqueran de Beaujeu 1551) to 1837, when the Haras Nationaux installed a model herd to encourage the production of horses in the Camargue for the light cavalry. The second period lasted until 1927, when the Haras finally stopped the production of these horses in the Camargue. The third period lasted until the extension of arable agriculture and the enclosures of the postwar years (c. 1948).

For the first period, as well as a number of short historical references dating back to 1550 (see Table 2), there is a detailed report on Camargue agriculture (Poulle 1817), including a section on horses. These writings give a reasonable picture of what the horses looked like, of the attempts to improve them, of the way in which they were managed, the uses to which they were put, and the environmental pressures on them. Most of the early references to these horses concern attempts to improve them by importing stallions from outside the Camargue. The Duke of Newcastle in 1660 warned the unwary of the practice of crossing local Camargue mares with Barb stallions, and then selling the products as imported Barbs. Shortly after this, in 1665, Colbert made an attempt to produce horses from this region for the light cavalry by placing imported African, Arab, and Barb stallions with local landowners, but the attempt was apparently a failure and had no lasting effects (Poulle 1817, Allier 1980). In 1727, Louis XIV had a stud established in the Camargue, which furnished horses for the royal stables, and the stud was renewed after the French Revolution by Napoleon I in 1806. However, Poulle (1817, section 230) states that the "defiant owners" have "always refused to use for reproduction, stallions which have not been produced by their local herds." He also points out that no selection of mares was practiced "[mares of] any shape whatsoever, even the most defective, are allowed to mate." The appearance of the horses described by Poulle (1817, Sections 218–223) differs little from that of present-day Camargues (see Table 1). He states that their hair color was dark when young and became uniformly white with age. Their height was about 1.38 m, rarely up to 1.49 m. They were hardy, and their conformation lacked the qualities of saddle horses.

The animals were maintained in relatively large herds of 20–60 animals, with generally only one adult stallion in order to avoid the "fights which would arise from rivalry and jealousy between stallions." Young males were, however, left in the herds, and though they were "pursued in vain by the adult males," they were "able, too often, to attract young females still of tender age . . . [and to achieve matings, which] led to degeneration and abortion rather than well-formed fruits" (Section 231). The herds grazed unimproved Camargue pastures throughout the year (Section 224), and though the burst of fresh plant growth in April quickly brought a shine to their coats and filled out the emaciated horses, their food became very scarce in January and "until the end of March the horses struggle incessantly against death." "We saw some which, with ravenous teeth, grazed the grass to the ground level, eating the earth attached to the roots and died from lourdige;[a] others which died simply from want and misery." (Section 224). The animals were infested by roundworms, probably strongylids and ascaroids. Occasionally, anthrax was recorded (Section 227).

Poulle notes that the horses were "tormented by the bites of insects . . . in the burning heat of summer" and lists *Stomoxys*, *Hippobosca*, Tabanidae, and Culicidae as the most abundant among them (Sections 79, 224).

The uses to which the horses were put in this period were, in order of importance, for threshing cereals, for riding ("gardians"[b] of the semiwild cattle and smallholders) and for draft. Occasionally, some were requisitioned or bought for cavalry work (the Protestant Jean Laporte in the 1660s; Napoleon in 1793–1794 and 1807). While none of these tasks was easy, the threshing was work which was apparently so punitive that only

---

[a] Lourdige is encephalitis, or the "staggers," per the author.
[b] Cattle herders.

Table 2. Historical Information

Authority	Date	Average shoulder ht (cm)	Coat color of adults	Hooves	Number of horses
de Quiqueran de Beaujeu	1551	—	—	—	4000
Roustan	1807	—	White, rarely other	Very hard	2000
de Truchet	1839	—	White	Small, round + very hard	—
de Rivière	1826	137	White, few grey	—	—
Poulle	1817	140	White	—	2000
Crespon	1844	—	White, rarely grey	—	—
Jessé de Charleval	1889	136	White	—	—
Pader	1890	133	White, rarely other	Strong, large, flat	450
Jacoulet and Chomel	1895	136	White, rarely bay, chestnut	—	800
Mathieu	1929	133	White, sometimes bay, black	Large, slightly flat	—
Aubert	1932	145	White, rarely other	Large, hard	—
Bérriot	1969	140	White	Broad + very hard	1200
Ministère de l'Agriculture	1978	135–145	White	Broad	2000

Camargue horses were capable of doing it. The season lasted 6–8 weeks in the summer, and the horses, often reproductive mares, were led from farm to farm so that they worked on most days. During the day's work, they covered about 80 km in the threshing arena, and they then, of course, had to try to satisfy their food requirements by feeding later.

Poulle's general conclusion is that the breed had, in spite of a physical "degeneration," "maintained these other qualities, in particular its endurance, without marked alterations; this must be attributed to the isolation of the country which hosts this breed, and to the defiance of its owners."

Forage		Biting insects	Climate	Other	Husbandry
Summer	Winter				
—	—	—	—	—	—
—	Very sparse	—	—	—	—
Aquatic	Perennials very sparse	Very common	Harsh	Poor water	None
—	—	—	—	—	—
Reeds etc. (good)	Grasses (very poor)	Very abun-dant	Harsh	—	None
Reeds (good)	Grasses coarse (starve)	—	—	—	—
—	—	—	—	—	—
Rich	—	—	—	—	None
Reeds (good)	Grasses (very sparse)	—	—	—	—
Good in marshes	Very sparse perennials	Very abun-dant	Cold winter	Water sparse spring	Shelter, coarse hay some herds
Reeds, rich	Perennials	Mosquitoes horseflies	Cold winter	—	None
Marsh (good)	—	Punitive	Harsh in winter + summer	—	Shelter, hay, foals brought in winter
—	—	—	—	—	—

(section 240). The preceding passages show that there were, in this domestic breed, strong selection pressures for efficient foraging behavior, and for adaptive responses to a variety of environmental pressures.

The second period, from 1837 to 1927, differs from the first, in that the attempts to cross-breed Camargue horses were successful. The hardy Camargue mares were used as brood mares to produce many thousands of cavalry horses, by mating them to stallions in the studs within and close to the Rhône delta. The stallions in the stud were principally Arabian, thoroughbred, and Breton horses. By 1881, 948 out of a population of

c. 1500 breeding mares were being mated by these foreign stallions (Allier 1980).

The uses to which the remaining horses of the Camargue were put at this time were the same as in the previous period, that is to say threshing, draft work and as riding horses for the gardians and farmers. Attempts were made to use the half-bred horses for these purposes, but it was found that only the so-called primitive Camargue type was hardy enough for the threshing. Breeding of pure Camargues therefore continued throughout this period, though it is certain that there was at least some gene flow from the foreign stallions, and it is probable that there was selection of mares for those with the best saddle-horse characteristics to cross with the foreign stallions. On the husbandry side, there were improvements in the pastures used by these mares, and some of them grazed in meadows and not simply in the salt flats and marshes of the Camargue. How many horses of the original Camargue breed remained at the end of this period is not known, but it is certain that there were only a few (Jacoulet and Chomel 1895), and even the gardians then rode larger, colored horses (Naudot 1977), which undoubtedly had superior qualities as saddle horses.[c]

After 1927, there was no further need for Camargue mares to produce cavalry horses, so the studs were removed, and the horses were once again used, as in the 18th century, for threshing, riding, and draft only. Husbandry was rudimentary, supplementary fodder was in general not given, nor was veterinary treatment. It is probable that disease, inclement weather, and the poor quality of the food exerted the same kind of environmental pressures on the horses during the years from 1927 to 1948 as it had done up until the 19th century.

During this period, most of the out-crossed horses must have disappeared, and the primitive type of Camargue horses maintained themselves because when the Camargue breed was officially recognized in 1978, the official description of the breed was little different from that of Poulle and other authors in the 150 years before; see Table 2.

Recent Trends

Since 1948, husbandry of Camargue horses has undergone several considerable changes; the main reason for this has been the appearance of

[c]Picon (1977, p. 71) quotes from an interview with a gardian in 1975; "Since there has been no cross-breeding for some years, we now have horses which are closer to the Camargue type than was the case before, and we have never heard so many complaints by riders that it is no longer possible to find a good horse. In the near future we shall have the true Camargue, and I wonder what we shall ride. . . . The Camargue makes a disagreeable mount." Translated by author.

tourism as a major factor in the local economy. Horses contribute to tourism in 2 ways: first, as vehicles in pony trekking, and second, as instruments for working the bulls, both in the traditional ferrades and in day-to-day management of the animals, most of which are used for the Provençal bull fights, (course à la cocarde). The importance of tourism in the area has meant that unlike other breeds, the number of Camargue horses has increased in the past century; see Table 2. Grazing areas, on the other hand, have declined, mainly because of the extension of arable agriculture, which followed the rice revolution of the postwar years. The density of the horses has been maintained by use of more intensive husbandry, veterinary treatment, and provision of supplementary fodder during the winter when food becomes limited.

The details of present-day husbandry vary a great deal from one herd to another. In a survey of management practices, herds were found which are managed under the traditional low-input system, as well as others which receive supplementary rations, are brought indoors for foaling, receive veterinary treatment and so on.

Management of the Tour du Valat Herd, 1948–1973

In 1947, the estate contained four mares. They were all of the primitive Camargue type and had been acquired between 1928 and 1938 from a neighboring owner.[d]

These mares were kept for the production of riding horses, and a herd log was maintained by F. Rensch and R. Lambert; the horses were allocated more extensive pastures, and their numbers built up to about 17 mares by the end of the 1950. The mortality rate was about 10% per annum (analysis of the herd log), and most of the excess animals were sold as foals, which were removed in their first autumn at age 8 months. Any remaining young males were removed, at ages 1–3 years, for training as riding horses. Some females were sold off at a variety of ages, but none was ever introduced from other breeding stocks.

Each year, a stallion was added to the herd for some months in the spring: between 1948 and 1973, 18 different breeding males were used, of which 8 came from other bloodlines, the remainder being young males which had not yet been removed for training.

The area of pasture devoted to the horses consisted of a number of fenced areas, and the horses were moved from one to another. In addition to the grazing in these pastures, the horses were occasionally put into the

---

[d]M. Cevoli (le Pape) of Port-St.-Louis, F. Rensch, personal communication, March, 1975.

Appendix 3   Climate

Temperatures

Monthly averages of the daily minimum and maximum temperatures are shown in Fig. 1. Air temperatures fell below zero in the months of December to February on average five times a month. In the period 1974–1983, snow was recorded twice, and the minimum temperature recorded was −5.5°C; the highest shade temperature was 33.6°C.

Wind

The area is well known for its high winds; only about 20% of the time is calm (Heurteaux 1975). The dominant sector is the northwest, from which comes the mistral winds. This sector accounts for 50–90% of the wind run, depending on the month: the average annual wind run is 149,000 km (1963–1973, Tour du Valat met. records) and wind speeds of over 100 km/hr occurred in 3–5 months of each year of the study, the highest recorded wind speed being 130 km/hr. High winds occurred in all months of the year, but the average wind run was lower in the summer than in the winter months; see Fig. 2. This seasonal difference will tend to exaggerate the impact of the climate on large mammals; there is less wind in the hot season and more in the cold, when the wind increases the animals' heat losses.

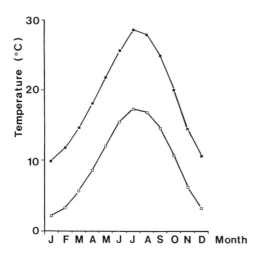

Fig. 1. Average daily maximum and minimum temperatures at Tour du Valat (1944–1973, data from Heurteaux 1975).

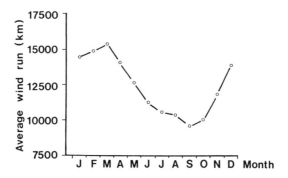

Fig. 2. Monthly averages of wind run at Tour du Valat (1963–1973, data from Heurteaux 1975).

Evaporation

Although the relative humidity is often high because of the proximity of the sea, the high summer temperatures lead to considerable rates of evaporation. The annual cycle of evaporation is shown in Fig. 3; the rates are lowest from November to January and highest in July. Though these measures of evaporation are not directly comparable with the rainfall, the summer rates of evaporation lead both to a deficit of water in the soil and to an arrest of plant growth outside the marshy habitats (Eckhardt 1972).

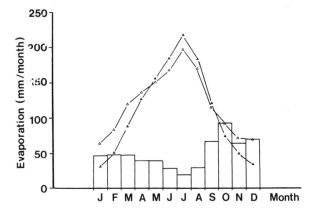

Fig. 3. Average monthly evaporation (triangles) and rainfall at Tour du Valat (histograms, 1944–1973); Piche evaporimeter (open triangles), Colorado evaporimeter (filled triangles, data from Heurteaux 1975).

Rainfall

The pattern of rainfall is unimodal, with about 60% falling between September and December; see Fig. 3. The summer months, especially July, are dry. Variability of rainfall has been studied on data collected at Salin de Giraud between 1888 and 1974 (Aris 1976); the coefficient of variation was relatively low for the months of September to May (0.79–0.97) but ranged between 0.97 and 1.46 in the summer months. The greatest variability therefore occurred during the period when the evaporation was greatest and the rainfall least.

Climatic Index

Emberger's (1955) index is

$$Q_2 = \frac{1000\,P}{M^2 - m^2}$$

where P is the annual precipitations in mm, M is the mean of the daily temperature maxima of the hottest month, m is the mean of the daily temperature minima of the coldest month.

The same method has been used for the years of this study because the application of more sophisticated indexes (e.g., Thornthwaite 1948), which take account of other variables such as wind run, did not seem warranted. However, it should be borne in mind that this index will probably underestimate evaporation in a windy environment such as the Camargue; for the data see page 29.

Appendix 4 Vegetation

A line transect was used at each of the 53 grid stakes (see Chapter 2, Fig. 2) to measure monthly the cover of aerial plant parts at 20 step-points. A sample of 28 stakes was chosen from these for more detailed measurements. Two plots were added to the sole stake that occurred in Facet 3, and marsh-edge plots (Group G) were grouped with Facet 2b; these are referred to as Facet 2 (see Table 1).

For logistic reasons, the methods used to measure the abundance and quality of the herbage were different in the higher ground and in the deep marsh.

Higher Ground

In Facets 2–7, samples of the herb layer (i.e., all nonwoody plants) were clipped to ground-level on four days around the 15th of each month. On the first occasion, a minimum of two rectangular (25 × 50 cm) quadrats were clipped and their corners marked with nails. A 2 × 1 m cage to exclude the mammalian grazers, covered on the vertical sides by a 2 × 2 cm wire mesh, was then placed over the quadrats. On the following 12 months, the "pre-clipped" (PC) quadrats were again clipped to ground level (PC samples), and two quadrats were again clipped from outside the cage ("OUT"). The cage was then placed over the sites of the OUT samples.

Table 1. Numbers of Vegetation Sampling Plots Allocated to Each Facet

Facet	Vegetation type	Number of plots
1	Deep marsh	5
2	Shallow marsh	4
F	Fallow fields	4
3	Wet enganes	3
4	Sansouïre	4
5	Enganes	3
6	Halophyte grassland	4
7	Coarse grassland	3

In Facets 3 and 5 (wet enganes and dry enganes), there were extensive areas of bare ground. The samples described previously were taken only from vegetated areas, and the plant biomasses were corrected for the proportion of bare ground, measured on step-point transects. No samples were clipped from the plots in Facet 4 (Sansouïre) because herbaceous plants were not found there.

The samples were brought back to the laboratory and dried immediately in an oven at 70°C for 24–48 hours until constant weights were obtained. One of each of the OUT samples was then sorted by hand into the components listed in Chapter 2 (quantity), and the weights of these five components were recorded. The other OUT sample was ground in a Wylie mill and analyzed both for the proportion of each plant species (see Chapter 4 and Appendix 6) and for the nutrients listed in Chapter 2.

Deep Marsh

The 2 m^2 in each cage were subdivided into a grid by wires, and all living aerial shoots were counted, and the heights of a sample of 25 were measured. The cage was closed, and a line transect, established close to the cage, was used to measure densities and heights outside. This was done by placing a wire quadrat (25 × 50 cm) at approximately 2-m intervals and counting the numbers of aerial shoots inside the quadrat. Living and dead shoots were recorded separately. The height of the shoot closest to the observer was also measured, one living and one dead for each species. Twenty-five such samples were taken at each of the five plots in the marsh.

The average height was then calculated for each species at each plot, and its weight was calculated, using the following equations. This weight was then multiplied by the estimated density (per/m^2), to give the estimated biomass for each species and plot.

A set of living aerial shoots of *Phragmites australis* and *Scirpus maritimus* was chosen by throwing a wire quadrat (25 × 50 cm) at chance in the deep marsh in August 1975. All living shoots inside the quadrats were clipped to ground level, dried at 70°C to constant weight, and weighed. Their heights were then measured to the nearest centimeter.

The relationships were nonlinear and are described by these equations:

$$\textit{Phragmites} \qquad y = 0.25\ e^{0.02x} \qquad r^2 = 0.94$$
$$\textit{Scirpus} \qquad y = 0.0013\ e^{1.66x} \qquad r^2 = 0.90$$

where y = shoot weight, and x = shoot height.

Appendix 5   Parasites

Internal Parasites

The horses are infected by eggs (ascaroids) or larvae (strongyloids). The larvae of the small strongyles settle in the mucosa of the wall of the large intestine, where they develop into adults, which remain attached to the same organ. Both larvae and adults feed on the mucosa.

The life cycle of the large nematodes involves more complex migrations: each species has a different route, which leads from the intestine immediately after infestion; through the blood system and such organs as the heart, pancreas, liver, and lungs (McGraw and Slocombe 1976); and into the large intestine, where the adult stages attach to the wall. The strongyles at this stage are 1.5–5 cm long and ascaroids 15–37 cm long. The lesions caused by the migrations of these parasites can be very damaging and sometimes cause the death of the host by blockages of the arteries (in particular the anterior mesenteric artery). The large nematodes are therefore potentially more dangerous than the small strongyles, which damage only the intestine wall.

Infestation in the Camargue herd was studied by J. Euzeby at the Ecole Nationale Veterinaire de Lyon, using coproscopy. As in other studies it was found that egg production by ascaroids began early in the lives of the foal and peaked in their first summer; see Fig. 1. Horses older than 2 years produced very few eggs, and it may be considered that they were practically free of these parasites.

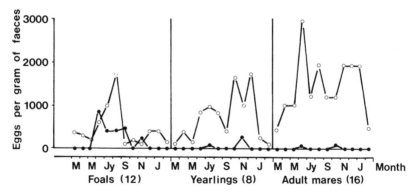

Fig. 1. Fecal egg counts in horses of different age classes; numbers of individuals in brackets: open circles – strongyloids; closed circles – ascaroids.

Table 1. The Composition by Genus of the Strongyle Eggs Excreted by Foals, and by Yearling Horses

Genus	Foals (1–12 months)	Yearlings (12–24 months)
Small strongyles		
Trichonema	92.2	93.8
Triodontophorus	1.33	0.75
Gyalocephalus	1.58	1.42
Poteriostomum	2.17	0.83
Oesophagodontus	1.5	0.58
Large strongyles		
Strongylus edentatus	0.58	1.50
Strongylus vulgaris	0.42	0.83
Strongylus equinus	0.25	0.25
	100.00	100.00

Strongyloid eggs included both large and small species but were dominated by the small strongyle *Trichonema*; see Table 1. The pattern of abundance followed a very different course from that of the ascaroids (see Fig. 1): peak production occurred between June and November in the different age classes, with lower levels being recorded in late winter. Egg production increased in older horses. Though egg production is not an accurate predictor of the worm burden (J.L. Duncan 1974), it can be assumed that the levels found in the adult mares correspond with a large population of worms, but no higher than have been found elsewhere (of the order of 10^3 eggs/g cf. J.L. Duncan 1974, Rubenstein and Hohmann 1989). Most of these were small strongyles, and thus the main effects of nematode parasites on the horses were (1) the direct consumption of tissue, and (2) decreased appetite, resulting from the damage to the intestine. Neither effect is easy to quantify, but it is likely that they were considerable.

The horses apparently develop an acquired immunity to ascaroids and large strongyles, which are potentially more damaging than the small strongyles. These larger parasites are virtually absent from horses older than 2 years.

External Parasites

The most important of these are the horseflies. Here, data are presented on the abundance of the blood-seeking females, and on their behavior toward the horses. This information is a necessary background to the studies described in Chapters 5 and 6; further details can be found in Hughes et al. (1981).

Temporal Variations in the Abundance of Different Species

Active females were sampled, using the Manitoba trap (Thorsteinson et al. 1965). It was sited near the study area and was operated throughout the seasons when flies were observed on the horses in the years 1976–1978, except for 1 week in May 1976.

The composition of the fauna (see Tables 2, 3, and 4) was similar to that reported by Raymond (1978), except for the inexplicable absence of

Table 2. Average Species Composition of the Total Tabanid Catch in a Manitoba Trap over Three Years (36962 individuals) and the Average Lengths of the Species

Abbreviation			% of catches 1976–1978	Average body length (mm)
HB	*Haematopota bigoti*	Gobert 1881	24.1	1.1 ± 0.02
HE	*Hybomitra explollicata*	(Pandellé 1883)	11.9	1.6 ± 0.02
HA	*H. acuminata*	(Loew 1858)	0.4	1.3 ± 0.02
HC	*H. ciureai*	(Séguy 1937)	8.6·	1.4 ± 0.03
TA	*Tabanus autumnalis*	L. 1761	16.9	1.9 ± 0.03
TB	*T. bromius*	L. 1758	27.3	1.3 ± 0.03
	T. bovinus	L. 1758	16 indivs	2.2 ± 0.04
TR	*T. rectus*	Loew 1858	15 indivs	2.3 ± 0.03
	T. eggeri	Schiner 1868	7 indivs	2.2
AF	*Atylotus flavoguttatus*	(Szilady 1915)	4.2	1.3 ± 0.02
AQ, AP	*A. quadrifarius*	(Loew 1874)	6.1	1.3 ± 0.03
C	⎰ *Chrysops caecutiens*	(L. 1758)	6 indivs	0.87
	⎱ *C. flavipes*	Meigen 1804	5 indivs	0.85
	C. viduatus	Fabr. 1794	0.1	0.90 ± 0.03

Note: Averages calculated with one standard error ($n = 10$).
The raw data are given in Table 3 (from Hughes et al. 1981, courtesy of CAB International).

Table 3. Composition of the Tabanid Fauna by Species (%) for the Years 1976–1978, by Week

Month	Week	HB			HE			AF			HC			TB		
		1976	1977	1978	1976	1977	1978	1976	1977	1978	1976	1977	1978	1976	1977	1978
May	1	—	—	—	—	3	0	—	1	0	—	—	—	—	—	—
	2	—	81	100	—	6	0	—	0	0	—	0	0	—	0	0
	3	—	81	89	—	56	0	—	6	0	3	2	3	—	0	0
	4	54	34	99	9	48	0	6	6	0	8	7	0	0	0	0
	5	42	19	88	8	16	1	12	17	0	6	16	3	3	3	0
June	1	49	35	64	16	51	0	6	15	1	10	10	12	3	3	2
	2	15	13	51	18	41	1	9	13	2	10	18	20	17	15	15
	3	4	2	26	12	26	0	6	16	1	9	13	27	33	43	40
	4	1	1	10	6	24	2	3	4	0	6	11	32	48	48	28
July	1	2	2	10	4	29	2	1	2	2	1	8	38	59	49	55
	2	1	2	2	1	8	1	0	0	1	1	5	27	74	68	70
	3	3	5	3	1	3	2	0	0	1	0	4	13	68	78	75
	4	0	5	2	0	3	2	0	0	0	1	1	12	45	68	78
August	1	1	2	3	1	3	2	0	1	0	0	2	4	60	72	62
	2	0	3	0	2	2	2	0	0	0	1	4	3	38	49	53
	3	1	0	5	4	1	1	1	0	0	3	3	1	24	23	17
	4	1	2	3	14	8	6	0	0	0	0	4	4	11	11	4
	5	10	2	11	11	18	0	0	0	0	1	5	4	8	8	4
September	1	42	9	43	8	9	0	0	0	0	0	0	4	19	3	6
	2	88	59	70	0	4	0	0	0	0	0	0	2	0	0	0
	3	90	70	97	1	0	0	0	0	0	0	0	0	2	0	0
	4	93	98	97	0	0	0	0	0	0	0	0	0	0	0	0
October	1	91	100	98	0	0	0	0	0	0	0	0	0	0	0	0
	2	98	100	100	0	0	0	0	0	0	0	0	0	0	0	0
	3	—	90	100	—	0	—	—	0	—	—	—	—	—	—	—
	4	—	—	—	—	—	—	—	—	—	—	—	—	—	—	—

Note: For abbreviations, see Table 2.

Table 4. Composition of the Tabanid Fauna by Species (%) cont., and Totals, for 1976–1978, by Week

Month	Week	TA 1976	TA 1977	TA 1978	AP 1976	AP 1977	AP 1978	Others 1976	Others 1977	Others 1978	n 1976	n 1977	n 1978
May	1	—	—	0	—	—	0	—	—	0	—	—	0
	2	—	15	8	—	0	0	—	0	0	—	74	1
	3	—	13	1	—	0	0	0	0	0	—	31	99
	4	28	2	9	0	0	0	1	0	1	1,799	139	203
	5	29	9	21	0	0	0	0	2	2	2,356	511	752
June	1	19	13	23	0	0	0	1	2	3	2,452	394	1,102
	2	28	8	27	2	0	0	1	0	1	3,109	897	645
	3	24	8	17	10	2	2	1	1	3	2,319	1,410	674
	4	17	11	15	16	3	3	0	0	1	2,267	1,018	366
July	1	14	10	10	13	4	5	1	1	0	1,123	1,735	265
	2	4	7	6	19	6	2	0	0	1	944	794	680
	3	9	8	6	16	5	4	2	1	1	693	441	589
	4	7	5	9	47	14	10	1	0	1	249	465	633
August	1	18	11	23	29	11	6	0	1	0	725	93	393
	2	28	9	31	32	27	6	0	0	3	323	148	202
	3	36	18	64	33	16	18	0	1	0	247	164	141
	4	41	48	63	30	33	1	1	0	0	157	61	141
	5	27	31	46	43	39	7	0	1	2	132	96	56
September	1	11	29	15	19	28	0	1	1	0	206	150	72
	2	3	6	2	9	30	0	0	0	1	33	215	122
	3	1	0	3	6	1	1	0	0	0	140	33	265
	4	3	1	2	4	0	0	0	0	0	67	664	125
October	1	0	0	0	9	0	0	0	0	0	92	208	64
	2	0	0	0	2	5	0	0	0	0	51	61	67
	3	—	0	—	—	—	—	—	0	—	—	19	0
	4	—	5	—	—	—	—	—	0	—	—	0	0
											19,484	9,821	7,657

Note: For abbreviations, see Table 2.

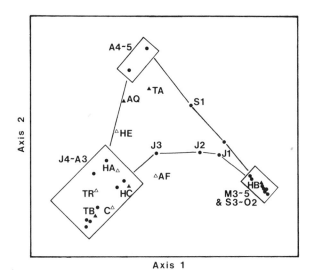

Fig. 2. Ordination of the weekly catches in the Manitoba trap during 1978. Weeks (circles) are numbered serially within months (e.g., A4–5 is the fourth and fifth week in August); tabanids are triangles (for abbreviations, see Table 2; filled triangles are species of particular importance in the construction of the axes).

Haematopota pluvialis (L. 1758), reported by Raymond from Tour du Valat. Closer comparisons with Raymond's results are not warranted because his sample of 1539 tabanids came from a large number of different localities within the delta. The flies ranged in size from the small *Chrysops* (0.9 cm, Table 2) to the large *Tabanus rectus* (2.3 cm).

There were marked changes in the composition of the fauna through the season. To analyze this, each species' contribution to the weekly catches was expressed as a percentage of the catch; the resulting Species × Weeks matrices was then subjected to correspondence analysis (Hill 1974). The first two axes of each year's analysis accounted for over 75% of the variance in the matrices, and the structure of the axes was similar in the three years. The ordination of the 1978 data is shown in Fig. 2. Axis 1 is bipolar and opposes two obvious groups of weeks: (1) a spring and autumn set, associated with the small *Haematopota* (Table 2); and (2) a summer set, associated with two medium-sized flies, *Hybomitra ciureai* and *Tabanus bromius*. Axis 2 opposes two late summer weeks (with *Tabanus autumnalis*, a large and abundant fly), to the rest of the data set.

The data on the percentage composition of the weekly catches were averaged over the 3 years, and the resulting matrix was analyzed by average linkage cluster analysis (Morgan et al. 1976). The average composition of the three seasonal groups is given in Fig. 3. As mentioned, the first

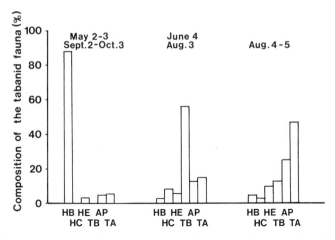

Fig. 3. Changes in the species composition of the tabanid fauna during the horsefly season. For abbreviations, see Table 2; May 2 is the second week of May.

group is dominated by a small species, *Haematopota* (1.1 cm), and the third by a large species, *T. autumnalis* (1.9 cm).

The reasons for these marked changes in the fauna are not known, but in studies of the responses of horses to the tabanids, account has been taken of them.

Spatial Variations in Abundance

It has been shown elsewhere that tabanids may show strong preferences for specific habitats (for four species, Jamnback and Wall 1959, pp. 33–46). To investigate whether this was the case in the study area, a riding horse was led to 53 numbered stakes, which were distributed systematically over the area (see Chapter 2, Fig. 2) and had been allocated to a land facet. This was done on 4 days of each month from May to September 1975, for between 10 and 17 hr. The observer allowed 10 minutes for tabanid numbers to stabilize and then counted the number settled on the horse. Four genera of tabanids were distinguished: *Tabanus*, *Haematopota*, *Hybomitra*, and *Atylotus*. The frequency distributions of the results were far from Gaussian and difficult to transform, so it was preferable to analyze the means rather than the raw observations. A three-way ANOVA without replication (Sokal and Rohlf 1972) showed (see Table 5) that, as could be expected from the results in the previous section, there were highly significant differences in the numbers of horseflies between months and between taxa, and the interaction between these factors was also significant, as the genera have different phenologies.

The effect of facet was also highly significant, with the fallow fields con-

Table 5. ANOVA (without replication) of the Effects of Taxon, Month, and Facet on the Numbers of Flies Settled on a Horse

Source	DF	SS	MS	F	p
Tabanid genus	3	48.7	16.23	23.65	0.001
Month	4	222.1	55.5	84.4	0.001
Facet	4	23.2	5.79	8.79	0.001
T × M	12	101.6	8.47	12.87	0.001
T × F	12	33.8	2.82	4.28	0.001
M × F	16	26.9	1.68	2.55	0.05
T × M × F	48	31.6	0.66	—	—

taining far more active flies (cf. Chapter 2, Fig. 19) than the other facets, between which there were no significant differences. The interaction Taxon–Facet was also highly significant; this arose from the fact that although all taxa showed the same trend (i.e., for there to be more individuals in the fallow fields), the effect was much more marked in the *Haematopota* genus than in the others. The interaction Month–Facet was weakly significant and arose principally from the deep marsh facet, which was relatively high in some months and low in others. The fallow fields were highest in all months.

In conclusion, the tabanids did show habitat preferences: the preferred habitat was the fields, and the preference was more marked in the small species (*Haematopota*). It is probable that the preference arises from the structure of the vegetation: the fields have not only a considerable herb layer (see Chapter 2, Fig. 9) but also well-developed hedges, bramble patches, and so on, and it is known that tabanids require vegetation both for protection against the wind and for resting in (Bailey 1948).

Variations in Numbers of Tabanids Attacking Individual Horses

Tabanids, like other blood-sucking Diptera, are known to prefer dark decoys to light-colored ones (Bracken et al. 1962). The effects of color of the horses on the number of tabanids attacking them were therefore tested. At the same time, the effects of sex, age, and social status of the horse were examined.

Counts of the number of tabanids attacking a horse were done by summing the number seen settled on the animal's skin with an estimate of the number circling it. These counts were easy to make in small herds (fewer than 10 horses) where the observer could move relatively freely. In larger herds, it was often not possible to distinguish the cloud of flies circling a particular horse because of the proximity of other animals. In this case, an index of abundance was used; 0 = none, 1 = few (1–3), 2 = some (4–10), 3 = many (11–30), 4 = very many (>30, Hughes et al. 1981), which could,

Table 6. The Number of Tabanids Attacking Horses of Different Sexes

Horse	Sex	Mean number of tabanids	Mean for the sex (*SE*)
1	♂	4.4	—
H1	♂	4.0	—
H4	♂	3.9	3.81 ± 0.31
I3	♂	4.1	—
I4	♂	2.6	—
5	♀	4.3	—
E1	♀	3.8	—
H2	♀	3.8	3.65 ± 0.23
I2	♀	3.6	—
J4	♀	2.9	—

Source: Data from Hughes et al. 1981.

if necessary, be converted to approximate numerical values by applying the geometric mean of each class (0, 2, 6, 18, 54).

A sample of five male and five female horses was chosen from the Tour du Valat herd, matched for age and for color. The number of tabanids settled on or circling these animals was recorded at hourly intervals during 24-hr watches. The results are given in Table 6 and show that there was no sex difference. There was, however, a suggestion that the numbers increased with age, and/or social status, which is correlated with age; this was strengthened by the fact that the oldest (20 year old) mare (No. 9) was found to be attacked by far more flies than any other horse.

In order to test whether this result was general, further observations were made on five neighboring herds whose members were approachable and which had at least five mares of similar size and color (white or near-white coats). These were visited on two successive days each, and 20 counts of tabanids were carried out as described previously, except that they were made at 5-minute intervals. The ages of the mares were known by their owners, and their dominance status was established either by field observations (cf. Clutton-Brock et al. 1976) or by inducing a high frequency of threats by providing a bucket of oats. It has been shown (Sereni and Bouissou 1978) that a hierarchy determined in this way is identical to one based on field observations.

The results were expressed as the percentage of the total herd count that was observed on each mare; see Fig. 4. It is clear that there is no consistent effect of age or dominance; however in each except Herd 3, there were one or two individuals which carried the major part of the fly burden. Considerable differences are known to exist among individual mammals in their attractiveness to biting insects such as mosquitoes; these are caused by individual (host) differences in the secretion of certain lipids, which attract the parasites. Whether this is true of the horse/tabanid interaction is not

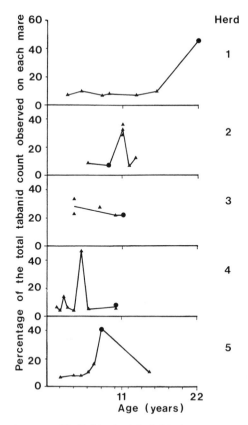

Fig. 4. The attractiveness of individual adult (white) mares to tabanids, in relation to their age (x axis) and status; dominant mares – black circles (results from five different herds; P. Cowtan, personal communication, August, 1977).

known, but the existence of these unusual individuals is an important feature of the interaction; it has been considered in the sampling programs.

The final characteristic of individuals which was studied was color. Because adult horses have a uniformly white coat color, it was necessary to use young animals (1–4 years), which provided a range of color from near black to near white. Three herds were used for these observations, two at Tour du Valat (D1 and G4) and one of the five mentioned above (Herd 4); each was observed over 2–3 days. The results are shown in Chapter 2, Fig. 20: it is clear that in each herd, there was a relationship between the color of a horse and the numbers of tabanids attacking it, with the darkest class being attacked by about three times as many flies as the palest.

It is therefore possible that the color (white) of the adults' coats is an adaptation which reduces biting fly attacks. The fact that Camargue cattle are black is in apparent contradiction of this idea; however, the much thicker skins of the cattle may be their primary defense, and the black color in these cattle may be linked to increased hardiness, as has been shown for cattle in the highland areas of Kenya (Finch and Western 1977).

There remains the possibility that the white coat color is an adaptation to minimize heat stress. However in Senegal, where ambient temperatures are much higher than in the Camargue, only 40% ($n = 286$) of the horses have white coats (personal observation, 1980). It therefore seems likely that the prevalence of white in Camargue horses is primarily an adaptation to reduce the numbers of horseflies attacking them.

Responses by the Horses to Attacks by Tabanids

Horses respond both indirectly, in their grouping patterns (Duncan and Vigne 1979), and directly, by "comfort behavior," to these attacks. Comfort movements include

Muscle twitching
Swishing the tail
Shaking the head, neck, and body
Striking and scratching with the muzzle
Stamping
Striking and scratching with a leg
Rolling

The frequency of such movements was measured for individual horses (Hughes et al. 1981) immediately after an estimate of fly abundance had been made. The frequency clearly increased as abundance increased, but nonlinearly (see Chapter 2, Fig. 21); presumably a degree of habituation occurred as the abundance of flies increased.

The frequency of responses increased from about 0 to nearly 1 per sec; when the flies became very abundant, comfort behavior was almost continuous. The frequency of comfort behavior is, therefore, a good index of the level of tabanid attacks: this was a useful result, as the close approach of an observer to count tabanids was undesirable for some studies (e.g., of social interactions).

The Dynamics of the Tabanid Load on Horse Herds

On four occasions during the summer of 1976, continuous 24-hr observations were made on both of the groups of horses that were then in the study areas, the breeding herd and the bachelor herd. The main aim of these observations was to study the horses' time budgets (see Chapter 5), but it was possible to study the dynamics of horsefly abundance on the herds at the same time.

Interpretation of the results was helped by the fact that a continuous record was kept of the location of the herds, and the activities (grazing, moving, resting) were also known.

A sample of 12 horses representing all age classes, and therefore colors, was selected in the breeding herd, and every 5 minutes, an estimate was made of the abundance index of tabanids around one horse. The sequence of horses observed was kept the same, so that the whole sample was covered every hour and a mean abundance index obtained. With the bachelor herd, observations were made on one of the horses every 5 minutes, and again, the sequence was kept constant. When these were from four males, this meant that the whole group was observed three times in an hour. When there were six, it was twice in an hour, but when eight were present the hourly relation was lost. The bachelor herd was a relatively uniform group, so it was still reasonable to express the results as an hourly mean value (see Chapter 2, Fig. 22).

Appendix 6　Methods Used for the Study of the Horses' Diets, Food Intake, and Digestion

Microhistological Analysis of Fecal Samples

A fecal sample was collected from every horse in the herds monthly or bimonthly during 1975–1976 by Steve Skelton, and thereafter from adults only. Samples weighing approximately 300 g were collected from fresh fecal deposits of individual horses and dried at 80°C to constant weight in an air-extraction oven. A subsample of several boluses was taken from the dried sample and ground in a Wylie mill equipped with a 1-mm grinding grid. The ground material was mixed with water and then washed in a 0.25-mm sieve. This yielded a fecal sample with a fairly consistent particle size range of 0.25–1 mm. An approximately 5-g subsample was taken from the sieve contents and treated according to the method of Stewart (1965). A reference collection was developed for plants known to occur in the study area using the techniques of Storr (1961). Black and white microphotographs of reference slides served as a key for identification of plant fragments in the fecal samples.

A Wild microscope equipped with a millimeter-calibrated stage and a Dynascope projection apparatus Was used for the analysis of fecal samples. A magnification of 200× was used in the analysis. Horizontal transects of the microscope slide were made at 2-mm intervals, using the calibrated stage (Field 1968). Consistent sample fragment size assured that no bias would be introduced through multiple "hits" on the same fragment;

the results thus provide an estimate of dietary proportions by volume. This is an accurate estimator of the proportions by weight for most plant mixtures (see Sparks and Malechek 1968, Holechek and Goss 1982).

Field's (1968) criteria were used to distinguish "identifiable" from "unidentifiable" fragments. Fragments were classed as "identifiable" if they intersected the center of the microscope field while making the transect, included some recognizable long cells or stomata or other features, and could be identified as belonging to a particular species. Hits on "identifiable" fragments were recorded until they totaled 100 for each individual horse sample. After the first 10 collection periods, only 50 "identifiable" fragments were recorded for each sample, to reduce the time required for analysis. No appreciable reduction in accuracy or precision was found as a result of the reduction in the number of counts per sample.

Two groups of species were difficult to distinguish and have been combined in the results: *Juncus subulatus* was combined with *J. gerardi*, and *Carex chaetophylla*, *C. vulpina*, and *Hordeum marinum* were combined.

The analysis for 1975–1976 was performed by S.T. Skelton (Skelton 1978) at Tour du Valat, and subsequent ones by the Composition Analysis Laboratory, University of Colorado, Fort Collins, Colorado 80523. There were no significant differences between observers for five duplicate samples analyzed by both laboratories ($p > 0.05$).

The accuracy of fecal analysis as a method of studying diet was studied by feeding hand-collected diets of known composition to stabled horses. These experiments are fully described in Skelton (1978); their number (4) was limited, but they concerned plants eaten by the free-ranging horses. In each case, the plant species made up similar proportions of diet and feces ($p > 0.05$, t-tests on arcsine-transformed percentages). The only exception was the white clover, *Trifolium repens*, which was significantly underrepresented in the feces. This was presumably because *Trifolium* contains less fiber and is more completely digested than the other species. The results for this species are therefore inaccurate; because it composed only a small part of the diet, no corrections have been made.

Seasonal Classification

The analysis of seasonal changes in diet was based on the results from the larger, breeding herd. The average percentage contribution by each plant species to the diet was calculated for each sampling occasion across all horses >2 years old. These averages were then subjected to single-linkage cluster analysis (Morgan et al. 1976). Four seasons were defined on the basis of the resulting dendrogram, and the data for both herds, transformed by arcsine, were subjected to a three-way ANOVA (season × herd × plant taxon), in order to test the statistical significance of the seasonal classification. Only the 14 species which composed >5% of the material in at least one sample were used.

Feeding Selectivity

Woody plants were not considered here, as they posed considerable sampling problems and because it was clear from prior observations, and from the literature, that horses avoided this category of plants.

Assessment of the proportions of different plant taxa in the sward was done by using the clipped samples described in Appendix 4. The monthly samples from each numbered stake were ground in the same way as the fecal samples. The material from each facet was then bulked, thoroughly mixed, and a subsample taken.

This was then treated exactly as the fecal samples, and the plant composition was determined and expressed as a proportion (S_i) for each taxon (i). The composition of the sward in Facet 1, the deep marsh, was determined indirectly, by height and density measurements, as described in Appendix 4. The proportion of the study area covered by each facet was known (see Chapter 2, Table 1), so the availability (A, %) of each taxon (i) was calculated by

$$A_i = \sum_{ij} (S_{ij} \times B_j) \times 100 \qquad (1)$$

where S_{ij} = proportion (0–1.0) of the sward in Facet$_j$ composed of Taxon$_i$; B_j = proportion (0–1.0) of the study area covered by Facet$_j$.

Selection (P) was assessed by the relation

$$P_i = \frac{D_i}{A_i} \qquad (2)$$

where D = percentage of the diet composed of Taxon$_i$, and A = availability (%) of Taxon$_i$, calculated from Equation 1.

The values of P (Equation 2) could vary between 0 and α, but in fact varied between 0 and 15; values >1.0 indicated selection, and values <1.0 avoidance of a taxon. The significance of departures from 1.0 were tested using χ^2 test, the expected number was taken as A_i and observed as D_i (i.e., $n = 100$) because the samples from different horses were not independent. The actual number of fragments examined was either 900 or 450.

Home Ranges and Proximity Relations of the Horses

The herds were visited regularly throughout the daytime (7–19 hr) and their location noted on the map shown in Chapter 2, Fig. 2. From these records, monthly and annual summaries were prepared. For the breeding herd, in addition, first and second nearest neighbors and their activities (grazing, resting, or other) were noted, as described in Wells and von Goldschmidt-Rothschild (1979).

Nutritional Value of the Diet

The crude protein, energy, calcium, phosphorous, and fiber contents of the plants were measured, using standard methods, as described in Chapter 2. This was done for all swards from March 1975 to June 1976. In addition, samples of the principal plant species grazed by the horses were also taken in 1979 during the fecal collection studies described below. These data covered the months January, February, March, and May. Ten samples were taken for each species, divided into two sets, and analyzed separately.

The fecal samples of individual horses collected in 1975–1976 were analyzed for nitrogen. In subsequent years (1977–1983), samples were taken monthly from reproductive females and every 3 months for males >3 years of age and analyzed in the same way. Fecal crude protein (CP_f) was estimated as $\%N \times 6.25$ (see also Salter and Hudson 1979).

Estimation of Dietary Crude Protein from the Fecal Crude Protein Concentration

The concentrations of dietary and fecal crude protein (CP_d, CP_f) are measured routinely in most digestibility trials, but for brevity, CP_f is seldom quoted.

Provided that the dietary crude protein concentration (CP_d) and the apparent digestibility coefficients of dry matter (DDM) and of crude protein (DCP) are reported, the results of digestibility trials can be used to calculate the fecal crude protein concentration (CP_f).

If a = amount of dry matter (DM)   ingested per day (kg)
  b = amount of crude protein (CP)   ingested per day (kg)
  x = amount of dry matter   excreted per day (kg)
  y = amount of crude protein   excreted per day (kg)

then, expressed fractionally, not as percentages,

$$DDM = \frac{a - x}{a}$$

$$DCP = \frac{b - y}{b}$$

$$CP_d = \frac{b}{a}$$

$$CP_f = \frac{y}{x}$$

By substitution,

$$CP_f = \frac{b(1 - DCP)}{a(1 - DDM)}$$

Rearranging,

$$CP_f = CP_d \cdot \frac{1 - DCP}{1 - DDM} \tag{3}$$

These data are available in a number of papers (Lindsay et al. 1926, Olsson et al. 1949, Fonnesbeck et al. 1967, Darlington and Hershberger 1968, Vander Noot and Gilbreath 1970, Vander Noot and Trout 1971, Hintz et al. 1972). In a few cases, the digestibility coefficient of organic matter (DOM) was reported, not DDM. The differences between these two coefficients are small but significant in horses: Where necessary, DDM has been predicted (from data in Fonnesbeck et al. 1967, Fonnesbeck 1968, 1969, and Vander Noot and Gilbreath 1970) by DDM = 1.042 DOM = 0.8775, $r = 30.99$, $n = 16$, $p < 0.001$).

The regression of CP_f on CP_d was analyzed by species (alfalfa *Medicago sativa*), orchard grass *Dactylis glomerata*, timothy *Phleum pratense*, and "other") and by trial.

Among species, alfalfa differed significantly from the others in slope ($b = 0.333$, $F = 8.30$, $df = 1, 28$; $p < 0.05$) and intercept ($a = 5.8$; $F = 11.9$; $df = 1, 28$; $p < 0.01$). The other species were homogeneous. The reason for this difference between alfalfa and the grasses is that the DCP of good-quality alfalfa is very much higher than DDM. On the limited data available, this does not seem to be true of the grasses; see Table 1. The two data points available for red clover suggested that this plant behaves like grasses. Because alfalfa is not usually part of the diet of free-ranging horses, this species has been left out of the remainder of this analysis.

Among trials, the results of Fonnesbeck et al. (1967), Experiment 2 differed from the others in intercept ($a = 5.18$; $F = 21.3$ $df = 1, 26$; $p < 0.001$).

Table 1. The Digestibility Coefficients of Crude Protein and of Dry Matter in Six High-Quality Forages with Protein $\geq 10\%$

	Digestibility of crude protein (%)		Digestibility of dry matter (%)	
	Alfalfa	Grasses	Alfalfa	Grasses
	75	60	61	50
	71	58	56	48
	65	62	52	51
	75	68	69	66
	72	67	62	63
	73	65	57	60
Mean	72	63	59.5	58

Source: Data from the references cited in the text.

---

[a] Only data from single, unground forages in these papers were used.

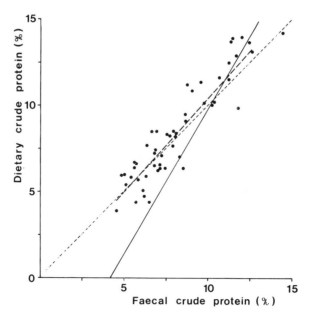

Fig. 1. The relationship between the crude protein concentration in the diets and in the feces of horses (equation 4, large dashes). The relationship in tropical cattle is also shown (solid line, $CP_d = 1.677\ CP_f - 6.93$, $r^2 = 0.92$; Bredon et al. 1963); the line of equality (small dashes) is also shown.

As a rule, CP_f is less than CP_d for a given forage because the digestibility of crude protein in equids generally exceeds that of dry matter for forages of this quality (8–12% CP_d). In this experiment, the reverse was true for four fifths of the forages. The reason for this is unclear, but I have preferred to exclude this trial from the remainder of this analysis.

The remaining results consist of 52 pairs of values for grass and clover diets, with CP_d ranging between 4 and 15%. The regression equation for the prediction of CP_d in grass and clover hays from CP_f is

$$CP_d = 1.09\ CP_f - 0.32 \qquad (4)$$
$$r = 0.936\ (r^2 = 0.88)$$

This regression may have a slightly lower slope than the equivalent one for grass hays and tropical cattle (Fig. 1, horses $b = 1.09$, cattle $b = 1.68$) and CP_f at $CP_d = 0$ is lower in horses than in cattle (0.29% vs. 4.1%). These differences are consistent with the fact that metabolic fecal protein per unit fecal dry matter is lower in horses than in cattle (Axelsson 1941).

The regression is unlikely to be valid for CP_d values outside this range: at high values, it would be expected to asymptote as DCP increases faster than DDM, with increasing CP_d. Below 5% CP_d, most of the protein in the feces will be of metabolic origin, and CP_f will tend to a constant value.

The Quantities of Food Eaten

The daily dry matter intake (DMI) was calculated from the relation

$$\text{DMI} = \text{fecal output} \times \left(\frac{100}{100 - \text{DMD}}\right) \tag{5}$$

where DMD is the digestibility of dry matter, estimated as described later in this section.

The fecal collections were made for a sample of reproductive mares (C1, D2, I2, I7), which were watched individually, initially for periods of 24 hr. It was later found that there was no difference in fecal output between night and day, so some mares were watched during the daytime only.

Fecal output (mean gDM/hr)

Horses	Month	Daytime	Nighttime
C1	Feb.	502	501
I7	May	238	237
D2	June	247	242

Their fecal output was measured by weighing each defecation on a Pesola field balance (± 25 g). The time of each defecation was noted. One in three was dried at 80°C to constant weight in the air-extraction oven, to determine dry matter content. Any exceptional samples, such as those that were urinated on by other horses, were also dried individually and treated separately for the estimation of dry fecal output per 24 hours. A composite sample of uncontaminated material was analyzed for nitrogen. Occasional samples were lost into the marshes: their weights were estimated from regressions developed for each mare and season, relating the dry weight of a defecation (in g) to the time elapsed (min) since the previous one ($b = 6.2 - 8.5$, $r^2 = 0.61 - 0.82$, all $p < 0.01$).

Daily dry matter fecal output was then calculated by summing the fresh weights of the dung, multiplying by the average proportion of dry matter, and adding to the result the individually dried exceptional samples and the calculated dry weights of any lost samples. The lost samples never-exceeded 10% of the total.

Body weights were estimated by one member of the team, J.C. Gleize. His estimates were calibrated when the horses were weighed in a Marechal

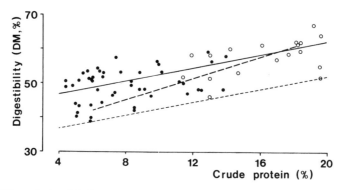

Fig. 2. The relationship between the digestibility of dry matter and its crude protein content. Data from legumes (open circles; mainly alfalfa) and grasses (filled circles). The equation (solid line) is given in the text (6). The minimum estimated is shown (small dashes). The regression of Vander Noot and Trout (1971) is also shown (large dashes).

balance (± 1 kg). For the range 350–460 kg, the regression of true on estimated weight had a slope (1.07) that was not significantly different from unity. The correlation was high ($r^2 = 0.90$), and no estimate was >20 kg from the true value.

Estimation of the Coefficients of Apparent Digestibility of Dry Matter (DDM) and of Crude Protein (DCP)[b]

The best predictor of DDM is the dietary concentration of NDF (neutral detergent fiber), but without fistulated animals, NDF in the diet cannot be measured. Dietary CP is the next best predictor and can be estimated in free-ranging animals (see preceding text).

The only published equations for the prediction of DDM and DCP for equids are those of Vander Noot and Trout (1971), based on four forages.

In order to broaden the base of these equations, the data from the papers listed in the section *Nutritional Value of the Diet* above were used. DOM values were used to calculate DDM, as described in cases where DDM values were unavailable.

For forages with CP_d of 4–20% (Fig. 2)

$$DDM = 43 + 0.96\ CP_d$$
$$r = 0.62,\ F_{1,79} = 48.7,\ p < 0.001 \tag{6}$$

---

[b] A review of prediction equations for DDM and CP_d from CP_f will shortly appear in *Grass and Forage Science*, by S. Boulot and W. Martin-Rosset (in press).

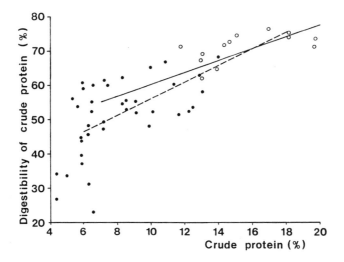

Fig. 3. The relationship between the digestibility of crude protein and the concentration of crude protein in the forage. Data from legumes (open circles; mainly alfalfa) and grasses (filled circles). The equation (solid line) is given in the text, equation 7. The regression of Vander Noot and Trout (1971) is also shown (dashed line).

For DCP, the relationship was curvilinear, with a turning point at about 7% CP_d. Above this value, DCP is given by Fig. 3

$$DCP = 43.3 + 1.72 \, CP_d \qquad (7)$$
$$r = 0.75, n = 31, p < 0.001$$

Below 7% CP_d, DCP may vary between 25 and 60% (median 45%).

These equations do not allow prediction of the digestion coefficients with accuracy, and there are two problems in the calculation of the errors on the estimates of DDM:

1. DDM, DCP, and CP_d are not independent
2. There is error in the measurement of CP_d

For these reasons, the 95% confidence limits have not been calculated. The maximum deviation below the regression line was 10 digestibility units for DDM amd 12.5 units for DCP. These deviations have been used as lower limits for the estimates.

Fecal crude protein was measured on three subsamples for each day using the micro-Kjeldahl technique. The digestibility of dry matter was calculated, using equations 4 and 6 above. Daily dry matter intake was then calculated from equation 5.

The Weight of Digesta in Horses and Cattle

Four Camargue cattle and four horses that were killed for other reasons were used to measure the weight of digesta. Their liveweight was measured to the nearest 1 kg before death. The GI tract was extracted and separated by string into the different compartments, which were then weighed separately, with and without their contents.

The contents of each compartment were then thoroughly mixed; two subsamples were taken and dried to constant weight at 80°C in the air-extraction oven.

Appendix 7 Methods Used for the Study of Social Behavior

Group Sizes and Home Ranges

From the beginning in 1974, the females asociated closely. They were considered to be in more than one group when the distance between the two (or more) groups was greater than the diameter of any group. Typical group size was calculated as for Jarman (1974) for females in their third year or older (i.e., potential breeders). All weaned horses were included in the measure of group size.

From July 1976–July 1978, the location of the horses was recorded on plastic overlays over c. 1:5,000 air photos. The numbered stakes were marked on the photos, and these together with vegetation patterns and individual bushes, which could be identified with ease, allowed the outlines of the herds to be drawn accurately. On these overlays, the activities of the horses (foraging, resting, traveling, other) were recorded, a note was made of the vegetation being used, and an estimate was made of the numbers of biting flies on a sample horse, using the abundance index described in Chapter 2, Fig. 21, and in Appendix 5 *Responses by the Horses.*

For the rest of the study, at hourly intervals during all fieldwork, the observer noted the location of the herd on a map of the type shown in Chapter 2, Fig. 2, with reference to vegetation boundaries and the numbered stakes, together with the information listed previously.

Dispersion, Social Behavior and Feeding Behavior of Individuals within the Groups

Dispersion and Social Behavior

For the period 1976–1978 the area occupied by the herd was calculated from the plastic overlays, using a dot-counting method, and from this was calculated the average area occupied by weaned horses. Foals were not included separately because their. mothers' tolerance was such that they did not at first occupy separate space.

During the observations of foraging behavior, which were described in Chapter 5, the facet in which the focal horse was standing at the beginning of each 5-min block was noted. The frequency of responses was recorded as an index of the abundance of horseflies (see Appendix 5 *Responses of the horses* as was the number of horses within 5 m of the focal animal. The head threats in which the animal was involved were recorded as they occurred, together with the identity of the partner.

The methods of recording interactions for the period 1974–1975 have been fully described in Wells and von Goldschmidt-Rothschild (1979). The results used here were obtained by following one to three focal animals, chosen without reference to their activity, and noting over a period of 15 min the occurrence of a number of behavioral interactions. All horses were watched in this way three times a day, on 4 days of 3 weeks in the months April–June and 4 weeks in November 1975. In this appendix, I consider the results for *head threats* (ears laid back and head extended toward another horse; mouth may be open in an attempted or successful bite) and for *friendly contacts* (all body contacts excluding play and agonistic interactions—the great majority were touching and rubbing aimed at removing flies). From 1979 to 1983, a reduced number of interactions was noted for the older horses (>2 years, about 1000 horse-hours of observation per year). Here, I present the results for head threats only: these allowed the rank of each animal to be determined each year from 1978 to 1983, using the method of Schein and Fohrman (1955).

In the analysis, behaviors were scored in bouts; a bout was terminated if a full minute passed without a particular behavior being directed at the same partner.

At hourly or half-hourly intervals, records were made of the nearest and second nearest neighbor of each horse, either by drawing a sketch map of the herd or by dictating the observations into a tape recorder.

Genetic Relatedness

For most of this study the herd was visited daily. This meant that all births were noted, including ones where the foals were stillborn or died at birth, for the mares carried blood on their tails and legs for at least a day. The mothers of all foals were therefore known.

To determine the fathers, in 1979, the herd was rounded up, and blood samples were taken from each horse. Blood types and serum protein types were determined for 28 loci by A.M. Scott. Paternity was determined by standard exclusion methods with all males >1 year old considered as potential sires. For the six foals where two or three horses could have been the sire, the most likely sire was determined by the likelihood method (Foltz and Hoogland 1981). This statistical method was checked against matings, which were recorded whenever they were seen, and against the identity of the dominant stallion of the mother's family unit at the time of conception, 11 months before the foal's birth. These sources of evidence were never contradictory in this study.

For horses conceived before the study, the father was taken to be the stallion present in the herd 335 days before the foal's birth. There was only ever one stallion, and the dates of its presence were recorded in the herd log (see Appendix 2). From these data, parentage could be determined for the whole herd, 1974–1979; see Appendix 8. Thereafter, a large proportion of foals had more than one potential sire because the number of stallions and their relatedness increased sharply.

Foraging Behavior in Herds of Different Sizes

During summer 1978, there were two small groups (I4, Bachelors) of 3 and 5 horses, which were generally isolated from the D group, a large herd of 47 horses.

In the small groups, there were three mares, I7, J3, and K4. The foraging behavior of these mares and of three mares in the D group, of similar age and reproductive status, was observed by Pamela Moncur, using the methods described in Chapter 5.

Adequate amounts of data were obtained for the principal facets (fields and deep marsh) only; the mares did not differ significantly within herd sizes, so the observations were pooled.

Appendix 8 Parentage of the Horses in the Tour Du Valat Herd 1974–1979

Horse	Sex	Mother	Father	Year of birth
9	F	X1	X2	1955
5	F	X1	X3	1961
7	F	9	X4	1964
C1	F	X5	X6	1968
D1	M	X7	X8	1969
D2	F	5	X9	1969
E1	F	7	X9	1970
G2	M	X5	X9	1972 (1976)
G3	F	X10	X9	1976
G4	M	9	X9	1976
H1	M	E1	X11	1973
H2	F	5	X11	1973
H3	F	7	X11	1973 (1975)
H4	M	9	X11	1975
I1	F	C1	X12	1974
I2	F	D2	X11	1974
I3	M	7	X12	1974
I4	M	5	X12	1974
I5	M	E1	X12	1974
I6	M	G3	X12	1974
I7	F	9	D1	1974
I8	F	C1	D1	1974
J1	M	7	D1	1975
J2	F	D2	D1	1975
J3	F	E1	G4	1975
J4	F	5	D1	1975
J5	F	G3	D1	1975
J6	F	9	D1	1975
J7	M	C1	G4	1975
K1[b]	M	7	G4	1976
K2	F	D2	D1	1976
K3	M	E1	G4	1976
K4	F	H2	D1	1976
K5	F	5	G4	1976
K6[b]	M	G3	G4	1976
K7	F	I1	G4	1976
K8	M	C1	D1	1976
L1[b]	F	7	H4	1977

Source: Data from Duncan et al. 1984a, courtesy of Academic Press
Note: Those horses that were not part of the study herd are indicated by a code name consisting of X + number. Year of death, where appropriate, is given in parentheses.
[a] Father–daughter mating.
[b] Mating of siblings not from same matriarchal family.

Horse	Sex	Mother	Father	Year of birth
L2	M	D2	H4	1977
L3	F	E1	D1	1977
L4	M	I2	I3	1977
L5	M	5	D1	1977
L6	F	G3	H4	1977
L7	M	H2	D1	1977
L8	M	9	D1	1977
M1	F	I1	G4	1978
M2	F	I8	H1	1978
M3	M	E1	D1	1978
M4[a]	F	J3	G4	1978
M5	M	J4	G4	1978
M6	M	J5	I3	1978
M7	F	5	D1	1978
M8	M	I7	H1	1978
M9	F	H2	G4	1978
M10[b]	M	I2	D1	1978
M11	M	G3	H1	1978
M12[b]	M	J6	G4	1978
M13	M	K2	G4	1978
M14	F	C1	D1	1978
N1	M	D2	H1	1979
N2	F	E1	H4	1979
N3[a]	M	K5	G4	1979
N4	M	I1	G4	1979
N5	F	J3	I4	1979
N6	M	I8	G4	1979
N7	F	9	H1	1979
N8	M	5	G4	1979
N9	F	K7	I3	1979
N10	M	J4	G4	1979
N11	M	J5	G4	1979
N12	M	I7	I4	1979
N13	M	I2	I3	1979
N14	F	H2	I5	1979
N15	F	G3	H1	1979
N16	F	L1	I4	1979
N17	M	7	H1	1979
N18	F	L3	I6	1979
N19	F	K2	H5	1979

Appendix 9   The Methods Used for the Study of Reproduction and Growth

Food Abundance and the Horses' Requirements

Data on the abundance of herbaceous plants were available (Chapter 2, page 37) for February 1975, and January of the years 1976–1977 and 1979–1985. The overall abundance was calculated as

$$\sum_{i=1}^{F} (B_i \times A_i)$$

where $i = 5, 6, 7$, and F, the principal winter feeding facets (Chapter 5); B_i is the mean oven-dry biomass of herbaceous plants (t/ha) in the "i"th facet; and A_i is the area of the "i"th facet (ha).

In Chapter 4, it was shown that in winter, the mares ate about 4.2% of their body weight in dry plant matter per day. Assuming that males and subadults eat as much, a very rough estimate of the herd's food requirement per day can be made by multiplying the measured biomass by 0.042. This measurement of food intake was made at a time (1979) when food was relatively abundant: in later years, there was less available, and the horses' intake may have changed, but it will give a rough estimate for the purposes of comparison with the quantities of food available.

Condition and Body Weights

Each horse was weighed individually in a Marechal balance (± 1 kg) twice in each ecological year (1.9–31.8) from 1979, between September 22 to 28 and March 18 to April 11. They were also weighed on December 1, 1982. The conceptus weight for pregnant females was estimated, using the method of Martin-Rosset and Doreau (1984a).

Body condition changed rapidly at some times, and it was not desirable to weigh the horses more often because of the disturbance this caused; a visual index of condition was therefore used (see Chapter 7, Table 2). This index was necessarily subjective, but the data were collected by the same two observers, so it is unlikely that major seasonal or annual trends were obscured. A record was made of the condition of each individual in January and May 1976–1977 and each month from October 1978 to August 1983.

Dates of Births and of Weaning

Birth dates were determined as described in Appendix 7. Sucklings were noted ad lib, which provided on average an observation per week; the weaning date was taken as the last observation of suckling plus 1 week.

Appendix 10   Methods Used for the Study of the Horses' Impact on the Vegetation

Permanent Exclosures

Five sites were chosen in 1975 as (a) representative of the major vegetation types of the pasture (Bassett 1978), (b) homogeneous, and (c) large enough to carry a set of nine study plots (7 × 7 m). The sites and exclosures are fully described in Bassett (1978); briefly, their characteristics were as follows.

Marshes

Site B. Deep marsh: three emergent species, dominated by *Scirpus maritimus* and some submerged *Characeae*. Maximum depth of water c. 1 m.
Site C. Shallow marsh: five species dominated by *Aeluropus littoralis*. Maximum depth of water, c. 50 cm.

Salt Flats

Site E. Enganes: dwarf scrub with eight species, dominated by *Arthrocnemum fruticosum*.

Grasslands

Site A. Halophyte grassland with about 40 species.
Site D. Halophyte grassland with about 50 species.

Site A was close to a large rabbit warren (15–50 m); D was slightly further from another large warren (50–150 m).

Two types of exclosure were built in 1976, one (X) fenced with 3-cm mesh, wire which successfully excluded rabbits, and one (Y) with two strands of barbed wire, which allowed free access to rabbits, coypu and boar, but excluded horses. A third type of plot (Z), marked with pegs at the corners, allowed free access to all herbivores. These plots were replicated three times at each site except E, where only one was set up. One of these plots is shown in the photographic plates.

Each year in May–June (1976–1983), when most plants were flowering, measurements of species frequency, the height of the vegetation, and the amount of bare ground (made by P. Bassett, M.C. and J.M. d'Herbès, and J. Jardel).

1. The rooted frequency of each species of higher plant was measured in a set of stratified random quadrats (15×15 cm) in each grazing treatment.
2. The height of the vegetation in the marshes was measured as the mean height of the tallest plant in each quadrat; in grassland sites, the height above the ground of the center of a square of expanded polystyrene (15×15 cm) was measured.

From 1981, measurements were made of the percentage contribution of each species to the herb layers using the point-quadrat method (Daget and Poissonnet, 1971). Two transects were placed diagonally across each plot and a needle lowered at 10 cm intervals. Each time a part of a plant touched the needle, the species of the part was identified and recorded as a separate touch. The contribution (c) of species "i" was then calculated as

$$C = \frac{t_i}{\sum\limits_{i=1}^{n} t_i} \times 100$$

where t is the number of touches of species "i."

The species diversity of a plot was calculated as the reciprocal Simpson Index

$$H_s = \frac{1}{\sum\limits_{i=1}^{n} p_i^2} \times 100$$

$$p_i = \frac{f_i}{\sum\limits_{i=1}^{n} f_i} \times 100$$

and f_i is the frequency of occurrence of the "i"th species in the 50 quadrats (15×15 cm).

Annual Exclosures

In the deep marsh, a set of six mobile exclosures (1×2 m) was set up in the deep marsh in February 1975 and 1979–1983. The height, density, and biomass of emergent plants were measured as described in Appendix 4 ("Deep Marsh" section) at the end of each growing season (August–September), inside and outside the exclosures. The peak biomass inside the exclosures is an index of ANPP (perhaps a half of the true value, Linthurst and Reimold 1978), and the difference between inside and outside gave a measure of the impact of grazing on plant structure over one season.

In addition to the emergent plants, submerged *Characeae* and *Ranunculus baudoti* grew in the deep marsh when water was present, usually September–June. Their peak biomass was measured in May 1975 and 1982–1983 by clipping 12 quadrats (25×50 cm) in the deep marsh.

Shrubs

Woodland (mainly elm) covered a very small part (<0.1%) of the area; unfortunately, dutch elm disease killed most of the trees in the early 1980s. The cover of shrubs (*Phillyrea angustifolia* and *Atriplex halimus*) was measured on aerial photographs from 1974 and 1980 by J.-M. d'Herbès, using a binocular microscope and a dot grid. Field checks showed that all bushes >40 cm in diameter were visible on the photographs, but many of the smaller ones were missed. The values of cover obtained by this method are therefore underestimates of the true values.

Primary Production and Offtake by the Horses

A rough estimate of primary production has been made on the basis of the data in Chapter 2, Table 4. This is here compared with the estimated offtake by the horses. Two independent estimates were made, one based on the direct measures made in 1979 on four mares; see Appendix 6 (section on "The Quantities of Food Eaten"). The other was obtained by calculating the energy requirements of the horses for maintenance, growth, activity, and reproduction, estimating the digestibility of energy in their food from the dietary crude protein concentrations, and thus estimating food intake.

Requirements of digestible energy for maintenance and growth of horses are given in NRC (1978). The values for horses of 425 kg mature weight (the average weight of adult horses) were obtained by interpolation. The requirements for reproduction in females are given in the same source and consist of 15.5 Mcal/day for late pregnancy and 25.2 Mcal/day for lactation. Requirements for activity have been estimated by a number of authors; I have used the values of Hintz et al. (1971). Time spent in each activity has been taken from Chapter 5, Table 2, for each age/sex class separately. No account has been taken of requirements for thermoregulation, so the values obtained for the winter will be underestimated, but probably not by much, as the temperatures were usually within the thermoneutral zone of cattle, and therefore probably of horses too (Senft and Rittenhouse 1985).

The horses lost weight in winter and gained in the warm seasons. In most years, gains balanced losses; no account has been taken of the energy losses involved in catabolizing and synthesizing these reserves.

To simplify the calculations, it was assumed that all pregnant females were in reproductive synchrony, with foaling occurring on April 15th. The year was divided into four equal seasons, in order to take account of differences in food quality: September–November (autumn), December–February (winter), March–May (spring) and June–August (summer). Reproductive females were taken as being in peak lactation in summer, half lactation in autumn, and late pregnancy in winter; and for the spring, half were taken as late pregnancy and half in peak lactation because the median foaling date occurred midway through the period.

The digestible energy in the food (Mcal/kg) was estimated from the equation

$$DE(Mcal/kg) = 1.510 + 0.0446 \, CP_d \tag{1}$$
$$r = 0.853, n = 42, p < 0.001$$

which was calculated from data on grasses and legumes in NRC (1978, Table 5). CP_d was calculated as described in Appendix 6 (section on *Nutritional Value of the Diet*).

References

Aguesse P (1960) Contribution a l'étude écologique des zygoptères de Camargue. Thèse doctorat, University of Paris

Albon SD, Mitchell B, Staines BW (1983) Fertility and body weight in female red deer: a density dependent relationship. J. Anim. Ecol. 52: 969–980

Alexander G (1974) The evolution of social behaviour. Ann. Rev. Ecol. Syst. 5: 325–383

Allden WG (1970) The effects of nutritional deprivation on the subsequent productivity of sheep and cattle. Nutr. Abstr. Rev. 40: 1167–1184

Allden WG, Whittaker IAM (1970) The determinants of herbage intake in grazing sheep: the interrelationships of factors influencing herbage intake and availability. Aust. J. Agr. Res. 2 1: 755–766

Allier P (1980) L'activité des Haras Nationaux en Camargue de 1750 à 1927. Courrier du Parc Nat. Rég. Camargue 18–19: 17–30

Allison T, Cicchetti DV (1976) Sleep in mammals: ecological and constitutional correlates. Science 194: 732–734

Allouche L, Tamisier A (1984) Feeding convergence of gadwall, coot and the other herbivorous waterfowl species wintering in the Camargue: a preliminary approach. Wildfowl 35: 135–142

Andrewartha HG, Birch LC (1984) The web of life. University of Chicago Press, Chicago

Aris H (1976) Etude statistique de données météorologiques journalières. Rapport technique No. 7609, Centre de Recherches Informatiques et Gestion, Université de Montpellier, France

Arnold GW (1963) Factors within plant associations affecting the behaviour and performance of grazing animals. In Crisp D (ed) Grazing in terrestrial and marine ecosystems. Br. Ecol. Soc. Symp. 4 Blackwells, Oxford p 133

Arnold GW, Dudzinski ML (1978) Ethology of free living domestic animals. Elsevier, Amsterdam

Arnold GW, Hill JL (1972) Chemical factors affecting selection of food plants by animals. In Harbourne JB (ed) Phytochemical ecology. Academic Press, London pp 71–101

Aubert FJ (1932) La race chevaline. Larguier, Nimes, France

Axelsson J (1941) Das Futterverdauungsvermogen des Pferdes. Sonderdruk aus Tierernahrung 13: 399–412

Baile CA (1979) Regulation of energy balance and control of food intake. In Church DC (ed) Digestive physiology and nutrition of ruminants. Vol 2 Chapter 11(2). O & B Books Inc., Corvallis, OR pp 291–320

Bailey NS (1948) The hovering and mating of Tabanidae: a review of the literature with some original observations. Ann. Ent. Soc. Amer. 41: 403–412

Bakker JP (1978) Changes in a saltmarsh vegetation as a result of grazing and mowing—five-year study of permanent plots. Vegetatio 38: 77–87

Bakker JP (1985) The impact of grazing on plant communities, plant populations and soil conditions on salt marshes. Vegetatio 62: 391–398

Bakker JP, Ruyter JC (1981) Effects of five years of grazing on a saltmarsh vegetation. Vegetatio 44: 81–100

Bassett PA (1978) The vegetation of a Camargue pasture. J. Ecol. 66: 803–827

Bassett PA (1980) Some effects of grazing on vegetation dynamics in the Camargue. Vegetatio 43: 173–184

Bauer K, Glutz (1966) Handbuch der Vogel Mitteleuropas. 6. Akademische Verkagsgesellschaft, Frankfurt/Main, Germany

Beeftink WG (1977) The structure of salt marsh communities in relation to environmental disturbances. In Jefferies RL, Dary AJ (eds) Ecological processes in coastal environments. Blackwell Scientific Publications, Oxford, England pp 77–93

Bell RHV (1969) The use of the herb layer by grazing ungulates in the Serengeti National Park, Tanzania. Ph.D. thesis, Manchester University, England

Bell RHV (1970) The use of the herb layer by grazing ungulates in the Serengeti. In Watson A (ed) Animal populations in relation to their food resources. Blackwell, Oxford, England pp 111–123

Bell RH V (1971) A grazing system in the Serengeti. Sci. Amer. 225: 86–93

Belovsky GE (1984) Herbivore optimal foraging: a comparative test of three models. Amer. Naturalist 124: 97–115

Belovsky GE (1986) Generalist herbivore foraging and its role in competitive interactions. Amer. Zool. 26: 51–69

Berger A, Corre JJ, Heim G (1978) Structure, productivité et régime hydrique de phytocénoses halophiles sous climat méditerrannéen. Terre Vie 32: 241–278

Berger J (1977) Organisational systems and dominance in feral horses in the Grand Canyon. Behav. Ecol. Sociobiol. 2: 131–146

Berger J (1983a) Induced abortion and soical factors in wild horses. Nature 303: 59–61

Berger J (1983b) Predation, sex ratios, and male competition in equids (Mammalia: Perissodactyla). J. Zool. Lond. 201: 205–216

Berger J (1986) Wild horses of the Great Basin. University of Chicago Press, Chicago

Berger J (1988) Social systems, resources and phylogenetic inertia: an experimental test and its limitations. In Slobodchikoff CN (ed) The ecology of social behaviour. Academic Press, London pp 157–188

Berger J, Cunningham C (1987) Influence of familiarity on the frequency of inbreeding in wild horses. Evolution 41(1): 229–231

Bérriot C (1969) Le cheval de Camargue. Thèse vétérinaire, Lyon, France

Bertram BCR (1978) Living in groups: predators and prey. In Krebs JR, Davies NB (eds) Behavioural ecology: an evolutionary approach (1st edn). Blackwells Scientific Publications, Oxford, England pp 64–96

Biber O (1975) Bibliographie de Camargue. C.R. activité de la Station Biologique de la Tour du Valat 16–53. Imp. C.G.C. Montpellier, France

Bideault G (1978) Le livre généalogique de la race de chevaux de Camargue. Courrier du Parc Nat. Rég. de Camargue. 14: 11–13

Bigot L (1961) Essai d'écologie quantitative sur les invertébres de la sansouire camarguaise. Thèse Doctorat d'Etat, Univ. d'Aix—Marseille, France

Blaxter KL, Wainman FW, Wilson RS (1961) The regulation of food intake by sheep. Anim. Prod. 3: 51

Blondel J, Isenmann P (1981) Guide des oiseaux de Camargue. Delachaux et Niestlé, Paris

Bongaarts J (1980) Does malnutrition affect fecundity? Science 208: 564–569

Boulot S (1987) L'ingestion chez la jument. Etude au cours du cycle gestationlactation: implications nutritionnelles et métaboliques. Thèse docteur ingenieur, E.N.S.A., Rennes

Boulot S (in press) Essai sur la Camargue. Actes Sud, Arles, France

Boulot S, Martin-Rosset W (in press) Evaluation of faecal indices to predict the quality of the diet of horses. Grass Forage Sci.

Bouman JG, Bos H (1979) Two symptoms of inbreeding depression in Przewalski horses living in captivity. In De Boer LE M, Bouman I, Bouman J (eds) Genetics and hereditary diseases of the Przewalski horse. Foundation for the Przewalski Horse, Rotterdam pp 111–117

Bourdon RM, Brinks JS (1982) Genetic, environmental and phenotypic relationships among gestation length, birth weight, growth traits and age at first calving in beef cattle. J. Anim. Sci. 55(3): 543–553

Bouwman H, Van der Schee W (1978) Composition and production of milk from Dutch warmblooded saddle horse mares. Z. Tierphysiol. Tierernahrg. u. Futtermittelkde. 40: 39–53

Box TW (1967) Influence of drought and grazing on mortality of five west Texas grasses. Ecology 48: 654–656

Boy V, Duncan P (1979) Time-budgets of Camargue horses: I. Developmental changes in the time-budgets of foals. Behaviour 71(3 + 4): 187–202

Boys H (1981) Food selection by some graminivorous Acrididae. Ph.D. thesis, Oxford, England

Bracken GK, Hanec W, Thorsteinson AJ (1962) The orientation of horse flies and deer flies (Tabanidae: Diptera): II. the role of some visual factors in the attractiveness of decoy silouhettes. Can. J. Zool. 40: 685–695

Braun-Blanquet J (1952) Les groupements végétaux de la France méditerranéenne. Centre National de la Recherche Scientifique, Paris

Bredon RM, Harker KW, Marshall B (1963) The nutritive value of grasses grown in Uganda when fed to zebu cattle: II. the relation between crude fibre and nitrogen free extract and other nutrients. J. Agr. Sci. 61: 105

Bredon RM, Torrell DT, Marshall B (1967) Measurement of selective grazing of tropical pastures using oesphageal fistulated steers. J. Range Manage. 20: 317

Bueno L, Ruckebusch Y, Dorchies P (1979) Disturbances of digestive motility in horses associated with strongyle infections. Vet. Parasitol. 5: 253–260

Campbell R (1930) Horses on the Camargue. In Adamastor. Faber and Faber, London

Campling RC (1970) Physical regulation of voluntary intake. In Phillipson AT (ed) Physiology of digestion and metabolism in the ruminant. Oriel Press, Newcastle-upon-Tyne, England

Campling RC, Lean IJ (1983) Food characteristics that limit voluntary intake. In Rook JA F, Thomas PC (eds) Nutritional physiology of farm animals. Longman, London pp 457–475

Carothers SW (1976) Feral asses on public lands: an analysis of biotic impact, legal considerations and management alternatives. Trans. North Am. Wildl. Nat. Resour. Conf. 41: 396–406

Carson K, Wood-Gush DG M (1983) Equine behaviour: II. A review of the literature on feeding, eliminative and resting behaviour. Appl. Anim. Ethol. 10: 179–190

Cary ER (1976) Territorial and reproductive behavior of the blackbuck antelope (*Antilope cervicapra*). Ph.D. thesis, Texas A & M University, College Station, Texas

Casebeer RL, Koss GG (1970) Food habits of wildebeest, zebra, hartebeest and cattle in Kenya Masailand. E. Afr. Wildl. J. 8: 25

Caughley G (1970a) Eruption of ungulate populations, with emphasis on the Himalayan thar in New Zealand. Ecology 51: 53–71

Caughley G (1970b) Fat reserves of Himalayan thar in New Zealand by season, sex area and age. New Zealand J. Sci. 13(2): 209–219

Caughley G (1977) Analysis of vertebrate populations. Wiley, London

Caughley G, Sheperd N, Short J (1987) Kangaroos: their ecology and management in the sheep rangelands of Australia. Cambridge University Press, Cambridge, England

Chacon E, Stobbs TH (1976) Influence of progressive defoliation of a grass sward on the eating behaviour of cattle. Aust. J. Agr. Res. 27: 709–727

Chenost M, Martin-Rosset W (1985) Comparaison entre espèces (mouton, cheval, bovin) de la digestibilité et des quantités ingérées des fourrages verts. Ann. Zootech. 34(3): 291–312

Church DC (1979) Digestive physiology and nutrition of ruminants. Vol 1 2nd ed. O & B Books; Inc., Corvallis, OR

Chvala M, Lyneborg L, Moucha J (1972) The horse flies of Europe. Entomological Society of Copenhagen, Copenhagen

Cifelli RL (1981) Patterns of evolution among the Artiodactyla and Perissodactyla (Mammalia). Evolution 35(3): 433–440

Clemens ET, Stevens CE (1980) A comparison of gastro-intestinal transit time in ten species of mammal. J. Agr. Sci. 94: 735–737

Clutton-Brock J (1981) Domesticated animals from early times. Heinemanns, London

Clutton-Brock TH (1988) Reproductive success: studies in individual variation in contrasting breeding systems. University of Chicago Press, Chicago

Clutton-Brock TH (1989) Mammalian mating systems. Proc. R. Soc. Lond. B 236: 339–372

Clutton-Brock TH, Greenwood PJ, Powell RP (1976) Ranks and relationships in highland ponies and highland cows. Z. Tierpsychol. 41: 202–216

Clutton-Brock TH, Guinness FE, Albon SD (1982) Red deer: behavior and ecology of two sexes. University of Chicago Press, Chicago

Clutton-Brock TH, Harvey PH (1983) The functional significance of variation in body size in mammals. In Eisenberg JF, Kleiman DG (ed) Advances in the study of mammalian behavior. Special publication no. 7 of the American Society of Mammalogists.

Coetzee CG (1969) The distribution of mammals in the Namib Desert and adjoining inland escarpment. Scient. Pap. Namib Desert Res. Stn. 40: 23–36

Conrad HR, Pratt AD, Hibbs JW (1964) Regulation of feed intake in dairy cows: I. Changes in importance of physical and physiological factors with increasing digestibility. J. Dairy Sci. 47: 54–62

Corre JJ (1975) Etude phyto-écologique des milieux littoraux salés en Languedoc et en Camargue. Vol I & 2 Doctoral dissertation, Université de Montpellier

Crawley MJ (1983) Herbivory: the dynamics of animal plant interactions. Blackwell Scientific Publications, Oxford, England

Crawley MJ, Gray AJ, Edwards PJ (1987) Colonisation, succession and stability. Blackwell Scientific Publications, Oxford, England

Crespon J (1844) La faune méridionale. Tome I et Tome II Ballivet et Fabre. Nîmes, France

Cumming DHM (1982) The influence of large herbivores on savanna structure in Africa. In Huntley BJ, Walker BH (eds) The ecology of tropical savannas. Vol 42 Ecological Studies, Springer, Berlin pp 217–245

Daget P, Poissonnet J (1971) Une méthode d'analyse phytologique des prairies. Critères d'application. Ann. Agron. 22(1): 5–41

Dallaire A (1974) Recherches sur l'alternance des états de veille et de sommeil chez les equidés. Memoire de Maitre ès Sciences Vétérinaires, Ecole Nationale Vétérinaire de Toulouse, France

Darby HC (1983) The Changing Fenland. Cambridge University Press, Cambridge, England

Dardaillon M (1986) Seasonal variations on habitat selection and spatial distribution of wild boar (Sus scrofa) in the Camargue, southern France. Behav. Processes 13: 251–268

Darlington JM, Hershberger TV (1968) Effect of forage maturity on digestibility, intake and nutritive value of alfalfa, timothy and orchardgrass by equine. J. Anim. Sci. 27: 1572–1576

Dauphine TC (1975) Kidney weight fluctuations affecting the kidney fat index in caribou. J. Wildl. Mgmt. 39: 379–386

de Quiqueran de Beaujeu P (1551) La Provence lovée. Translated by de Claret F In Le Musée. Revue Arlesienne historique et littéraire. No. 14 (2nd series, 1875), pp 109–111

de Rivière M (1826) Mémoire sur la Camargue. Madame Huzard, libraire, 7 rue de l'Eperon Saint André, Paris

de Truchet M (1839) Mémoire sur les chevaux de Camargue. Garcins, Arles, France

Demarquilly C, Jarrige R (1974) The comparative nutritive value of grasses and legumes. Vaxtodling 28: 33–41

Demment MW, Greenwood GB (1988) Forage ingestion: effects of sward characteristics and body size. J. Anim. Sci. 66: 2380–2392

Demment MW, Van Soest PJ (1985) A nutritional explanation for body-size patterns of ruminant and nonruminant herbivores. Amer. Naturalist 125: 641–672

Dewsbury DA (1982) Dominance, rank, copulatory behaviour and differential reproduction. Q. Rev. Biol. 57: 135–159

Dittmer HJ (1973) Clipping effects on bermuda grass biomass. Ecology 54: 217–219

Dixon GM, Campbell AG (1978) Relationships between grazing animals and populations of the pasture insects Costelytra zealandica (White) and Inopus rubriceps (Macquart). N. Z. J. Agric. Res. 21: 301–305

Doreau M, Martin-Rosset W, Petit D (1980) Activités alimentaires nocturnes du cheval au pâturage. Ann. Zootech. 29(3): 299–304

Duboul-Razavet C (1955) Contribution a l'étude géologique et sédimentologique du delta du Rhone. Doctoral dissertation, University of Paris

Dudzinski ML, Schuth HJ, Wilcox DG, Gardiner HG, Morrissey JG (1978) Statistical and probabilistic estimators of forage conditions from grazing behaviour of merino sheep in a semi-arid environment. Appl. Anim. Ethol. 4: 357–368

Dunbar RIM (1978) Sexual behaviour and social relationships among gelada baboons. Anim. Behav. 26: 167–178

Dunbar RIM, Dunbar EP (1974) Social organisation and ecology of the klipspringer (*Oreotragus oreotragus*). Z. Tierpsychol. 35: 481–493

Duncan JL (1974) Field studies on the epidemiology of mixed strongyle infection in the horse. Vet. Rec. 94: 337–345

Duncan JL, Pirie HM (1974) The pathogenesis of single experimental infections with *Strongylus vulgaris* in foals. Res. Vet. Sci. 18: 82–93

Duncan P (1975) Topi and their food supply. Ph.D. thesis, Nairobi University, Kenya

Duncan P (1982) Foal killing by stallions. Appl. Anim. Ethol. 8: 567–570

Duncan P (1983) Determinants of the use of habitat by horses in a Mediterranean wetland. J. Anim. Ecol. 52: 93–111

Duncan P (1985) Time budgets of Camargue horses. III. Environmental influences. Behaviour 92(1–2): 188–208

Duncan P, Cowtan P (1980) An unusual choice of habitat helps horses avoid biting flies. Biol. Behav. 5: 55–60

Duncan P, D'Herbès JM (1982) The use of domestic herbivores in the management of wetlands for waterbirds in the Camargue, S. France. In Scott DA (ed) Managing wetlands and their birds. International Waterfowl Research Bureau, Slimbridge, Glos., England

Duncan P, Feh C, Malkas P, Gleize JC, Scott AM (1984a) Reduction of inbreeding in a natural herd of horses. Anim. Behav. 32(2): 520–528

Duncan P, Foose TJ, Gordon IJ, Gakahu CG, Lloyd M (1990) Comparative nutrient extraction from forages by grazing bovids and equids: a test of the nutritional model of equid/bovid competition and coexistence. Oecologia 84: 411–418

Duncan P, Vigne N (1979) The effect of group size in horses on the rate of attacks by blood-sucking flies. Anim. Behav. 27: 623–625

Eberhardt LL, Majorowicz AK, Wilcox JA (1982) Apparent rates of increase for two feral horse herds. J. Wildl. Manage. 46(2): 367–374

Ebhardt H (1954) Verhaltensweisen von Islandpferden in einem norddeutschen Freigelande. Saugetierkundl. Mitt. 2: 145–154

Eckardt FE (1972) Dynamique de l'écosysteme, stratégie des végétaux et échanges gazeux: cas des enganes à *Salicornia fruticosa*. Oecol. Plant. 7 333–345

Eisenmann V, Turlot JC (1978) Sur la taxinomié du genre Equus: description et discrimination des espèces d'après les données craniomètriques. Les Cahiers d'Analyse des Données Vol. III. No. 2: 127–130

Eline JF, Keiper RR (1979) Use of exclusion cages to study grazing effects on dune vegetation on Assateague island, Maryland. Proceedings of the Pennsylvania Academy of Science 53: 143–144

Emberger L (1955) Une classification biogéographique de climats. Recueil de Trav. des Labo. de Bot., Géol., Zool. de la Fac. des Sci. de l'Univ. de Montpellier, Bot. 7: 3–43

Evans DG (1977) The interpretation and analysis of subjective body condition scores. Anim. Prod. 26(2): 119–126

Evans JW, Borton A, Hintz HF, Van Vleck D (1977) The horse. Freeman, San Francisco, CA

Fedigan LM (1983) Dominance and reproductive success in primates. Anthropologie 26: 91–129

Feh C (1987) Etude du developpement des relations sociales chez des étalons de race Camargue et de leur contribution à l'organisation sociale du groupe. Diplôme de recherche Universitaire. Université d'Aix-Marseille II, France.

Feh C (1990) Long term paternity data in relation to different rank-aspects for Camargue stallions. Anim. Behav. 40(5): 995–996

Feist JD, McCullough DR (1976) Behaviour patterns and communication in feral horses. Z. Tierpsychol. 41: 337–371

Ferrazzini S (1987) Organisation dans le temps et l'espace des bovins de race Camarguaise. Plasticité du comportement dans un environnement hétérogène. Thèse doctorat, Univ. Rennes I, France

Field CR (1968) The food habits of some wild ungulates in Uganda. Ph.D. thesis, University of Cambridge, England

Field CR (1972) Food habits of wild ungulates in Uganda by analysis of stomach contents. East Afr. Wildl. J. 10: 17

Finch V, Western D (1977) Cattle colours in pastoral herds: natural selection or social preference. Ecology 58(6): 1384–1392

Fishbein W, Gutwein BM (1977) Paradoxical sleep and memory storage processes. Behav. Biology 425–464

Fletcher DJ C., Mitchener CD (eds) (1987) Recognition in animals. Wiley, New York

Foil LD (1989) Tabanids as vectors of disease agents. Parasitol. Today 5(3): 88–96

Foltz DW, Hoogland JL (1981) Analysis of the mating system in the black-tailed prairie dog (Cynomys ludovicanus) by likelihood of paternity. J. Mammal. 62: 706–711

Fonnesbeck PV (1968) Digestion of soluble and fibrous carbohydrates of forage by horses. J. Anim. Sci. 27: 1306–1344

Fonnesbeck PV (1969) Partitioning the nutrients of forages for horses. J. Anim. Sci. 28: 624–663

Fonnesbeck PV, Lydman RK, Vander Noot GW, Symons LD (1967) Digestibility of the proximate nutrients of forage by horses. J. Anim. Sci. 26: 1039–1045

Foose TJ (1978) Digestion in wild species of ruminant versus nonruminant ungulates. American Association of Zoological Parks and Aquaria Conf. Proc. 74–84

Foose TJ (1982) Trophic strategies of ruminant versus nonruminant ungulates. Ph.D. thesis, University of Chicago

Gakahu CG (1982) Feeding strategies of the Plains zebra, Equus quagga bohmi, in the Amboseli ecosystem. Ph.D. thesis, University of Nairobi, Kenya

Ganskopp D, Vavra M (1986) Habitat use by feral horses in a northern sagebrush steppe. J. Range Manage. 39(3): 207–212

George M, Ryder OA (1986) Mitochondrial DNA evolution in the Genus Equus. Mol. Biol. Evol. 3(6): 535–546

Ginsberg JR (1988) Social organization and mating strategies of an arid adapted equid: the Grevy's zebra. Ph.D thesis, Princeton, NJ

Gjersing FM (1975) Waterfowl production in relation to rest–rotation grazing. J. Range Manage. 28: 37–42

Gogan PJ P. (1973) Some aspects of nutrient utilisation by Burchell's zebra (Equus burchelli bohmi Matschie) in the Serengeti–Mara region, East Africa. M.Sc. thesis, Texas A & M University, College Station, Texas

Good JE G, Bryant R, Carlill P (1990) Distribution, longevity and survival of upland hawthorn (Crataegus monogyna) scrub in north Wales in relation to sheep grazing. J. Appl. Ecol. 27: 272–283

Gordon IJ (1989a) Vegetation community selection by ungulates on the Isle of Rhum. I. Food supply. J. Appl. Ecol. 26: 35–51

Gordon IJ (1989b) Vegetation community selection by ungulates on the Isle of Rhum: II. Vegetation community selection. J. Appl. Ecol. 26: 53–64

Gordon IJ (1989c) Vegetation community selection by ungulates in the Isle of Rhum: III. Determinants of vegetation community selection. J. Appl. Ecol. 26: 65–79

Gordon IJ, Duncan P (1988) Pastures new for conservation. New Scientist 1604: 54–59

Gordon IJ, Duncan P, Grillas P, Lecomte T (1990) The use of domestic herbivores in the conservation of the biological richness of European wetlands. Bulletin d'Ecologie. t. 21(3): 49–60

Gordon IJ, Illius AW (1988) Incisor arcade structure and diet selection in ruminants. Functional Ecology 2: 15–22

Gosling LM (1986) The evolution of mating strategies in male antelopes. In Rubenstein DA, Wrangham RW (eds) Ecological aspects of social evolution. Princeton University Press, Princeton pp 244–281

Grafen A (1990) Do animals really recognize kin? Anim. Behav. 39(1): 42–54

Granval P, Aliaga P, Soto P (1988) Quantified appreciation of land parcels characteristics in a wet zone as basis of the rural valorisation. In Lefeuvre JC (ed) Conservation and development: the sustainable use of wetland resources. Proceedings of the Third International Wetlands Conference, Sept. 19–23, Rennes, France. Museum National d'Histoire Naturelle, Paris pp 233–234

Greenfield SB, Smith D (1974) Diurnal variations of non-structural carbohydrates in the individual parts of switchgrass shoots at anthesis. J. Range Manage. 27(6): 466–469

Grime JP (1977) Evidence for the existence of three primary strategies in plants, and its relevance to ecological and evolutionary theory. Amer. Naturalist 111(982): 1169–1194

Grimsdell JJ (1973) Age determination of the African buffalo, *Synercerus caffer* Sparrman. E. Afr. Wildl. J. 11: 31–54

Groves CP (1974) Horses, asses and zebras in the wild. David & Charles, Newton Abbot, London

Grubb PJ (1977) The maintenance of species-richness in plant communities: the importance of the regeneration niche. Biol. Rev. 52: 107–145

Grubb P, Jewell PA (1974) Movement, daily activity and home range of Soay sheep. In Jewell PA, Milner C, Morton-Boyd J (eds) Island survivors: the ecology of the Soay sheep of St. Kilda. Athlone Press, University of London pp 160–194

Grzimek B (1952) Versuche über das Farbsehen von Pflanzenessern: I. Das farbige Sehen (und das Sehschärfe) von Pferden. Z. Tierpsych. 9: 23–39

Guthrie RD (1984) Mosaics, allelochemics and nutrients. An ecological theory of late Pleistocene. In Martin PS, Klein RG (eds) Quaternary extinctions. University of Arizona Press, Tucson pp 259–298

Gwynne MD, Bell RH V (1968) Selection of vegetation components by grazing ungulates in the Serengeti National Park. Nature 220: 390–393

Hacker JB (ed) (1982) Nutritional limits to animal production from pastures. Commonwealth Agricultural Bureaux, Farnham Royal, U.K.

Hafner H (1977) Contribution à l'étude écologique de quatre espèces de hérons (Egretta g. garzetta L., Ardeola r. ralloides Scop., Ardeola i. ibis L., Nycticorax n. nycticorax L.) pendant leur nidification en Camargue. Thèse Universitē, Toulouse, France

Haenlein GFW, Holdren RD, Yoon YM (1966a) Comparative response of horses and sheep to different physical forms of alfalfa hay. J. Anim. Sci. 25: 740–744

Haenlein GFW, Smith RC, Yoon YM (1966b) Determination of the fecal excretion rate of horses with chromic oxide. J. Anim. Sci. 25: 1091–1095

Hamilton BA, Hutchinson KJ, Annis PC, Donnelly JB (1973) Relationships between the diet selected by grazing sheep and the herbage on offer. Austr. J. Agric. Res. 24(2): 271–277

Hanley TA, Brady WW (1977) Feral burro impact on a Sonoran desert range. J. Range Manage. 30: 374–377

Hansen RM (1976) Foods of free-roaming horses in southern New Mexico. J. Range Mgmt. 29: 347–348

Hansen RM, Clark RC (1977) Foods of elk and other ungulates at low elevation in northwestern Colorado. J. Wildl. Manage. 41(1): 76–80

Harper JL (1969) The role of predation in vegetational diversity. In Woodwell GM, Smith HH (eds) Diversity and stability in ecological systems. Brookhaven National Laboratory, Upton, NY

Harper JL (1977) Population biology of plants. Academic Press, London

Harris AH, Porter LS W (1980) Late Pleistocene horses of Dry Cave, Eddy County, New Mexico. J. Mamm. 61(1): 46–65

Hausfater G, Hrdy SB (1984) Infanticide: comparative and evolutionary perspectives. Aldine, Hawthorne, NY

Hawkes JH, M, Daniluk P, Hintz HF, Schryver HF (1985) Feed preferences of ponies. Equine Vet. J. 17(1): 20–22

Hendrichs H, Hendrichs U (1971) Dikdik und Elefanten: Okologie und Soziologie zweler afrikanischer Huftiere. Piper and Co., Munich

Heurteaux P (1969) Recherches sur les rapports des eaux souterraines avec les eaux de surface (étangs, marais, rizières), les sols halomorphes et la végétation en Camargue. Thèse d'Etat, University of Montpellier, France

Heurteaux P (1970) Rapports des eaux souterraines avec les sols halomorphes et la végétation en Camargue. Terre Vie 24: 467–510

Heurteaux P (1975) Climatologie des années 1972 et 1973 en moyenne Camargue. Terre Vie 29: 151–160

Hik DS, Jeffreys RL (1990) Increases in the net above-ground primary production of a salt-marsh forage grass: a test of the predictions of the herbivore optimisation model. J. Ecol. 78(1): 180–195

Hill MO (1974) Correspondence analysis: a neglected multivariate method. Applied Statistics 23: 340–354

Hill MO, Bunce RG H, Shaw MW (1975) Indicator species analysis, a divisive polythetic method of classification, and its application to a survey of native pinewoods in Scotland. J. Ecol. 63: 597–613

Hintz HF (1969) Review article: comparison of digestion coefficients obtained with cattle, sheep, rabbits and horses. Veterinarian 6: 45–55

Hintz HF, Hintz RL, Van Vleck LD (1979) Growth rate of thoroughbreds. Effect of age of dam, year and month of birth, and sex of foal. J. Anim. Sci. 48(3): 480–487

Hintz HF, Loy RG (1966) Effects of pelleting on the nutritive value of horse rations. J. Anim. Sci. 25: 1059–1062

Hintz HF, Roberts SJ, Sapin SW, Schryver HF (1971) Energy requirements of light horses for various activities. J. Anim. Sci. 32(1): 100–106

Hintz HF, Schryver HF, Halbert M (1972) A note on the comparison of digestion by llama, guanaco, sheep and ponies. Anim. Prod. 16: 303–305

Hirst SM (1975) Ungulate habitat relationships in a South African savanna. Wildl. Monogr. No. 44: 1–60

Hodgson J, Wilkinson A (1967) The relationship between liveweight and herbage intake in grazing cattle. Anim. Prod. 9: 365–376

Hoffmann L (1958) An ecological sketch of the Camargue. Brit. Birds 51: 312–349

Hofmann RR (1973) The ruminant stomach. East Afr. Mongr. in Biology No. 2. East African Literature Bureau, Nairobi, Kenya

Hofmann RR (1989) Evolutionary steps of ecophysiological adaptation and diversification in ruminants. Oecologia 78: 443–457

Holechek JL, Goss BD (1982) Evaluation of different calculation procedures for microhistological analysis. J. Range Manage. 35(6): 721–723

Holechek JL, Valdez R, Schemnitz SD, Pieper RD, Davis CA (1982b) Manipulation of grazing to improve or maintain wildlife habitat. Wildl. Soc. Bull. 10: 204–210

Holechek JL, Vavra M, Pieper RD (1982a) Botanical composition of range herbi-
vores diets: a review. J. Range Manage. 35(1): 309–317
Houpt KA, Keiper R (1982) The position of the stallion in the equine dominance
hierarchy of feral and domestic ponies. J. Anim. Sci. 54(5): 945–950
Houpt KA, Law K, Martinisi V. (1978) Dominance hierarchies in domestic horses.
Appl. Anim. Ethol. 4: 273–283
Hughes AL (1988) The Quagga case: molecular evolution of an extinct species.
TREE 3(5): 95–96
Hughes RD, Duncan P, Dawson J (1981) Interactions between Camargue horses
and horseflies (Tabanidae). Bull. Ent. Res. 71: 227–242
Hunter RF (1962) Hill sheep and their pasture: a study of sheep grazing in south
east Scotland. J. Ecol. 50: 651–680
Hutchinson KJ, King KL (1980) The effects of sheep stocking level on invertebrate
abundance, biomass and energy utilisation in a temperate, sown grassland. J.
Appl. Ecol. 17: 369–387
I.U.C.N. (1988) Red list of threatened animals. The World Conservation Monitor-
ing Centre, Cambridge, England
Illius AW, Gordon IJ (1987) The allometry of food intake in grazing ruminants. J.
Anim. Ecol. 56: 989–999
Illius AW, Gordon IJ (1990) Constraints on diet selection and foraging behaviour
in mammalian herbivores. In Hughes RN (ed) Behavioural mechanisms of food
selection. Springer-Verlag, Berlin pp 369–392
Ivins JD (1952) The relative palatability of herbage plants. J. Brit. Grassld. Soc. 7:
43
Jacoulet J, Chomel C (1895) Traité de Hippologie. Gendron, Saumur, France
Jamnback H, Wall W (1959) The common salt-marsh Tabanidae of Long Island,
New York. NY State Museum and Science Service Bulletin No. 375: 1–77
Janis CM (1976) The evolutionary strategy of the Equidae and the origins of rumen
and cecal digestion. Evolution 30: 757–774
Janis CM (1989) A climatic explanation for patterns of evolutionary diversity in
ungulate mammals. Palaeontology (London) 32: 463–481
Janis CM, Gordon IJ, Illius AW (in press) Modelling equid/ruminant competition:
what happened to the North American browsing equids? Historical Biology
Janzen DH (1981) *Enterolobium cyclocarpum* seed passage rate and survival in
horses, Costa Rican Pleistocene seed dispersal agents. Ecology 62(3): 593–601
Janzen DH (1982) Differential seed survival and passage rates in cows and horses,
surrogate Pleistocene dispersal agents. Oikos 38: 150–156
Jarman MV, Jarman PJ (1973) Daily activity of impala. E. Afr. Wildl. J. 11: 75–92
Jarman PJ (1974) The social organisation of antelope in relation to their ecology.
Behaviour 48(3–4): 215–266
Jarman PJ, Sinclair ARE (1979) Feeding strategy and the pattern of resource-
partitioning in ungulates. In Sinclair ARE, Norton-Griffiths M (eds) Serengeti,
dynamics of an ecosystem. University of Chicago Press, Chicago pp 130–163
Jarrige R (ed) (1980) Alimentation des ruminants. Institut National de la Recher-
che Agronomique Publications, Versailles, France
Jessé de Charleval M (1889) Préliminaire d'une étude sur la Camargue. Imprimerie
Marseillaise, Marseille, France
Jewell PA, Holt S (eds) (1981) Problems in management of locally abundant wild
mammals. Academic Press, New York
Johnson GE, Borman MM, Rittenhouse LR (1982) Intake, apparent utilisation
and rate of digestion in mares and cows. Proc. Western Section, Amer. Soc.
Anim. Sci. 33: 3
Joubert E (1972) The social organization and associated behaviour in the Hart-
mann zebra (*Equus zebra hartmannae*). Madoqua Series I No. 6: 17–56

Joubert E (1974) Composition and limiting factors of a Khomas Hochland population of Hartmann's zebra (*Equus zebra hartmannae*). Madoqua Series I No. 8: 49–50

Joubert E, Louw GN (1977) Preliminary observations on the digestive and renal efficiency of Hartmann's zebra, *Equus zebra hartmannae*. Madoqua 10(2): 119–121

Kaiser PH, Berlinger SS, Fredrickson LH (1979) Response of blue-winged teal to range management on waterfowl production areas of south-eastern South Dakota. J. Range Manage. 32: 259–61

Kalinowska A, Mochnacka-Lawacz (1976) Grassland ecosystem. International Biological Programme summary sheet analysis. Pop. Ecol. Stud. 2(2): 163–169

Kaminski M, Duncan P (1981) Hemotypes and genetic structure of Camargue horses. Biochem. Syst. Ecol. 9(4): 365–371

Kaseda Y, Nozawa K, Mogi K (1984) Separation and independence of offsprings from the harem groups in Misaki horses. Jpn. J. Zootechnical Sci. 55(11): 852–857

Keiper RR (1979) Population dynamics of feral ponies. Proceedings of Symposium on the Ecology and Behavior of Wild and Feral Equids. Sept. 6–8, Laramie, University of Wyoming pp 175–183

Keiper RR (1980) Effect of management on the behavior of feral Assateague Island ponies. Proceedings of the Second Conference on Scientific Research in the National Parks. Vol. 8: 382–393

Keiper RR (1981) Ecological impact and carrying capacity of ponies. U.S. Fish and Wildlife Service, Chincoteague Nat. Wildl. Refuge, VA pp 1–20

Keiper RR, Berger J (1982) Refuge-seeking and pest avoidance by feral horses in desert and island environments. Appl. Anim. Ethol. 9: 11–120

Kerbes RH, Kotanen PM, Jefferies RL (1990) Destruction of wetland habitats by lesser snow geese: a keystone species on the west coast of Hudson Bay. J. Appl. Ecol. 27: 242–258

King KL, Hutchinson KJ (1976) The effects of sheep stocking intensity on the abundance and distribution of mesofauna in soils of pastures. J. Appl. Ecol. 13(1): 41–56

Kingdon J (1980) East African mammals. Vol. IIIC, IIID Academic Press, London

Klingel H (1967) Sociale Organisation und Verhalten Freilebender Steppenzebras (*Equus quagga*). Z. Tierpsychol. 24: 580–624

Klingel H (1968) Soziale organisation und Verhaltensweisen von Hartmann und Bergzebras (*Equus zebra hartmannae* und *E. z. zebra*). Z. Tierpsychol. 25: 76–88

Klingel H (1969) Social organisation and population ecology of the Plains zebra (*Equus quagga*). Zool. Africana 42(2): 249–263

Klingel H (1972) Social behaviour of African Equidae. Zoologica Africana 7: 175–185

Klingel H (1975) Social organisation and reproduction in equids. J. Reprod. Fert. Suppl. 23: 7–11

Klingel H (1977) Observations on social organization and behavior of African and Asiatic wild asses (*Equus africanus* and *E. hemionus*). Z. Tierpsychol. 44: 323–331

Kohli E (1980) 24-Stunden-Aktivitatszyklus freilebender Nutrias (Myocastor coypus) in der Camargue (Sudfrankreich). Z. F. Tierpsychologie 54(4): 368–380

Kohli E (1981) Untersuchungen zum Einfluss de Nutrias (*Myocastor coypus*) auf die naturliche Vegetation der Camargue. Ph.D. dissertation—naturwissenschaftlichen, Fakultat der Universitat Bern, Switzerland

Koller BL, Hintz HF, Robertson JB, Van Soest PJ (1978) Comparative cell wall and dry matter digestion in the cecum of the pony and the rumen of the cow using in vitro and nylon bag techniques. J. Anim. Sci. 47(1): 209–215

Kortlandt A (1984) Vegetation research and the "bulldozer" herbivores of tropical Africa. In Chadwick AC, Sutton CL (eds) Tropical rainforest. Special publication of the Leeds Philosophical and Literary Society, pp 205–226

Kownacki M, Sasimowski E, Budzynski M, Jezierski T, Kapron M, Jelen B, Jaworowska M, Dziedzic R, Seweryn A, Solmka Z (1978) Observations of the 24-hour rhythm of natural behaviour of Polish primitive horse bred for conservation of genetic resources in a forest reserve. Genetica Polonica 19: 61–77

Kozlowski J, Weigert RG (1986) Optimal allocation of energy to growth and reproduction. Theor. Pop. Biol. 29(1): 16–37

Krebs JR (1978) Optimal foraging decision rules for predators. In Krebs JR, Davies NR (eds) Behavioural ecology. Blackwell, Oxford pp 23–63

Krebs JR, Harvey PH (1986) Busy doing nothing—efficiently. Nature 320: 18–19

Kreulen DA (1985) Lick use by large herbivores: a review of benefits and banes of soil consumption. Mammal Rev. 15(3): 107–123

Kruit J (1955) Sediments of the Rhone delta: grain size and microfauna. Verh. Koninklijk-Mijnbouwkundig Genootschap, Geol. Ser. 15: 357–514

Kruuk H (1972) The spotted hyaena: a study of predation and social behavior. University of Chicago Press, Chicago

Krysl LJ, Hubbert ME, Sowell BE, Plumb GE, Jewett TK, Smith MA, Waggoner JW (1984) Horses and cattle grazing in the Wyoming Red Desert: 1. food habits and dietary overlap. J. Range Manage. 37(1): 72–76

Kubiena WL (1953) The soils of Europe. Murphy, London

Kvet J, Husak S (1978) Primary data on biomass and production estimates in typical stands of fispond littoral plant communities. In Dykyjova D, Kvet J (eds) Pond littoral ecosystems. Ecological Studies 28, Springer Verlag, Berlin pp 211–216

Lamprey HK (1963) Ecological separation of large mammal species in Tarangire Game Reserve, Tanganyika. E. Afr. Wildl. J. 1: 1

Langer P (1987) Evolutionary patterns of Perissodactyla and Artiodactyla (Mammalia) with different types of digestion. Z. Syst. Evolution-forsch 25: 212–236

Laurie WA, Brown D (1990) Population biology of marine iguanas (*Amblyrhynchus cristatus*): II. Changes in annual survival rates and the effects of size, sex age and fecundity in a population crash. J. Anim. Ecol. 59(2): 529–544

Laut JE, Houpt KA, Hintz HF, Houpt T (1985) The effect of caloric dilution on meal patterns and food intake of ponies. Physiol. Behav. 35: 549–554

Laws RM, Parker IS C, Johnstone RC B (1975) Elephants and their habitats: the ecology of elephants in North Bunyoro, Uganda. Clarendon Press, Oxford

Leader-Williams N, Ricketts C (1982) Seasonal and sexual patterns of growth and condition of reindeer introduced into South Georgia. Oikos 38: 27–39

Lecomte T, LeNeveu C (1986) Le Marais Vernier: contribution a l'etude et a la gestion d'une zone humide. Thèse doctoral, Universite de Rouen, France

Lecomte T, LeNeveu C, Jauneau A (1981) Restauration de biocenoses palustres par l'utilisation d'une race bovine ancienne (Highland cattle): cas de la Reserve Naturelle des Mannevilles (Marais Vernier—Eure). Bull. Ecol. 12(2/3): 225–247

Lemaire S, Tamisier A, Gagnier F (1987) Surface, distribution et diversité des principaux milieux de Camargue. Leur évolution de 1942 à 1984. Rev. Ecol. Terre Vie Suppl. 4: 47–56

Lenarz MS (1985) Lack of diet segregation between sexes and age groups in feral horses. Can. J. Zool. 63: 2583–2585

LeNeindre P, Petit M (1975) Nombre de tétées et temps de pâturage des veaux dans les troupeaux de vaches allaitantes. Ann. Zootech. 24(3): 553–558

Lindsay JB, Beals CL, Archibald JC (1926) The digestibility and energy value of feeds for horses. J. Agric. Res. 32: 569–604

Linthurst RA, Reimold RJ (1978) An evaluation of methods for estimating the net aerial primary productivity of estuarine angiosperms. J. Appl. Ecol. 15: 919–931

Lossaint P, Rapp M (1971) Répartition de la matière organique, productivité et cycle d'éléments mineraux dans les écosystèmes du climat méditerranéen. UNESCO Ecol. and Cons. 4: 597–617

Lowe VP W (1969) Population dynamics of the red deer (Cervus elaphus L.) on Rhum. J. Anim. Ecol. 38: 425–457

Lowman BG, Scott NA, Somerville SH (1976) Condition scoring of cattle. The East of Scotland College of Agriculture Bulletin. 6: 1–31

Mabutt JA (1968) Review of concepts of land classification. In Stewart GA (ed) Land evaluation. MacMillan, Melbourne, Australia

MacArthur RH, Levins R (1967) The limiting similarity, convergence and divergence of coexisting species. Amer. Naturalist 101: 377–385

MacFadden BJ (1976) Cladistic analysis of primitive equids. Systemat. Zool. 25(1): 1–14

MacFadden BJ (1977) "Eohippus" to Equus: fossil horses at the Yale Peabody Museum. Discovery 12(2): 69–76

Maddock L (1979) The "migration" and grazing succession. In Sinclair ARE, Norton-Griffiths M (eds) Serengeti: dynamics of an ecosystem. University of Chicago, Chicago pp 104–129

Maloiy GM O, Taylor CR, Clemens ET (1978) A comparison of gastrointestinal water-content and osmolarity in East African herbivores during hydration and dehydration. J. Agr. Sci., Camb. 91: 249–252

Mangel M (1990) Resource divisibility, predation and group formation. Anim. Behav. 39: 1163–1172

Margaris NS (1976) Structure and dynamics in a phyrganic (East Mediterranean) ecosystem. J. Biogeogr. 3(3): 249–259

Martin LD, Neuner AM (1978) The end of the Pleistocene in North America. Trans. Neb. Acad. Sci. 6: 117–126

Martin PS (1970) Pleistocene niches for alien animals. Bioscience 20: 218–221

Martin-Rosset W (1983) Revue bibliographique. Particularités de la croissance et du développement du cheval. Ann. Zootech. 32(1): 109–130

Martin-Rosset W, Andrieu J, Vermorel M, Dulphy J-P (1984) Valeur nutritive des aliments pour le cheval. In Jarrige R, Martin-Rosset W (eds) Le cheval: reproduction, sélection, alimentation, exploitation. Institut National de la Recherche Agronomique, Paris pp 209–238

Martin-Rosset W, Doreau M (1984a) Besoins et alimentation de la jument. In Jarrige R, Martin-Rosset W (eds) Le cheval: reproduction, sélection, alimentation, exploitation. Institut National de la Recherche Agronomique, Paris pp 355–370

Martin-Rosset W, Doreau M (1984b) Consommation d'aliments et de l'eau. In Jarrige R, Martin-Rosset W (eds) Le cheval: reproduction, sélection, alimentation, exploitation. Institut National de la Recherche Agronomique, Paris pp 333–354

Martin-Rosset W, Palmer E (1977) Bilan de trois années de monte en liberté dans un troupeau de juments de trait. Technical Bulletin C.R.Z.V. Theix I.N.R.A. 28: 33–39

Mathieu G (1929) Le cheval camargue. Son élevage, son amélioration. Thèse vétérinaire. Bosc Frères et Riou, Lyon, France

Mayes E, Duncan P (1986) The temporal patterns of foraging behaviour in free-ranging horses. Behaviour 96(1–2): 105–129

McCort W (1979) The feral asses (*Equus asinus*) of Ossabaw Island, Georgia: mating system and the effects of vasectomies as a population control procedure. In the proceedings of the Symposium on the Ecology and Behaviour of Wild and Feral Equids. Sept. 6–8 University of Wyoming, Laramie pp 71–83

McGraw BM, Slocombe JOD (1976) *Strongylus vulgaris* in the horse: a review. Can. Vet. J. 17: 150–157

McNaughton SJ (1979) Grassland–herbivore dynamics. In Siclair AR E, Norton-Griffiths M (eds) Serengeti: dynamics of an ecosystem. University of Chicago Press, Chicago pp 46–81

McNaughton SJ (1983) Compensatory growth as a response to herbivory. Oikos 40: 329–336

McNaughton SJ (1984) Grazing lawns: Animals in herds, plant form and coevolution. Amer. Naturalist 124:

McNaughton SJ, Sabuni GA (1988) Large African mammals as regulators of vegetation structure. In Werger MJA., Van der Aart PJM, During HJ, Verhoeven JTA (eds) Plant form and vegetation structure. SPB Academic Publishing, The Hague, Netherlands pp 339–354

Meddis R (1975) On the function of sleep. Anim. Behav. 23: 676–691

Mengin D (1980) Zoonoses bactériennes et virales en Camargue. Doctoral dissertation, Université de Montpellier, Montpellier, France

Metz JHM (1975) Time patterns of feeding and ruminating in domestic cattle. Mededelingen Landbouwhogeschool Wageningen (Netherlands) 75(12): 1–66

Meyer H, Alhswede L, Reinhardt HJ (1975) Untersuchungen uber Fressdauer, Kaufrequenz und Futterzerkleinerung beim Pferd Dtsch. Tierarztl. Wshr. 82: 54–58

Milchunas DG, Sala OE, Laurenroth WK (1988) A generalised model of the effects of grazing by large herbivores on grassland community structure. Amer. Naturalist 132: 87–106

Miller R, Denniston II RH (1979) Interband dominance in feral horses. Z. Tierpsychol. 51: 41–47

Milner C, Hughes RE (1968) Methods for the measurement of the primary productivity of grassland. IBP Handbook No. 6 Blackwell Scientific Publications, Oxford

Ministère de l'Agriculture. (1978) Arrêté relatif au Cheval Camargue. Journal officiel, Paris, France

Mitchell B, McCowan D, Nicholson IA (1976) Annual cycles of body weight and condition in Scottish red deer, *Cervus elaphus*. J. Zool., Lond. 180: 107–127

Moehlman P (1974) Behavior and ecology of feral asses (*Equus asinus*). Ph.D. thesis, University of Wisconsin, Madison

Moehlman P (1986) Ecology of cooperation in canids. In Rubenstein DI, Wrangham RW (eds) Ecological aspects of social evolution. Princeton University Press, Princeton, NJ pp 64–86

Moir RJ (1968) Handbook of physiology, section 6. (Alimentary canal Vol V Ch.126) American Physiology Society, Washington

Molinier R, Devaux JP (1978) Carte phytosociologique de la Camargue au 1/50.000ème. Biologie et Ecologie Méditerranéenne. V(4): 159

Molinier R, Tallon G (1970) Prodrome des unités phytosociologiques observées en Camargue. Bull. Mus. Hist. Nat. Marseille 29: 5–23

Molinier R, Tallon G (1974) Documents pour un inventaire des plantes vasculaires de la Camargue. Iere partie. Bull. Mus. Hist. Nat. Marseille 34: 7–165

Monard AM (1984) La tétée et le jeu: une étude des différences quantitatives entre poulains mâles et femelles de race Camargue. D.E.A. d'Ecologie/Ethologie, Univ. Rennes I, France

Moore CWE (1964) Distribution of grasslands. In Barnard C (ed) Grasses and grasslands. Macmillan, London pp 182–205

Morgan BJT, Simpson MJA, Hanby JP, Hall-Craggs J (1976) Theory and application of cluster analysis. Behaviour 56: 1–43

Morris MG, Plant R (1983) Reponses of grassland invertebrates to management by cutting: V. changes in Hemiptera following cessation of cutting. J. Appl. Ecol. 20: 157–177

Mungall EC (1979) Habitat preferences of Africa's recent Equidae, with special reference to the extinct quagga. In proceedings of the Symposium on the Ecology and Behavior of Wild and Feral Equids. Sept. 6–8 University of Wyoming, Laramie pp 159–172

National Research Council (1978) Nutrient requirements of domestic animals. Horses. National Academy Press, Washington DC

National Research Council (1982) Phase III: final report of committee on wild and free-roaming horses and burros to the board on agriculture and renewable resources commission on natural resources. National Academy Press, Washington DC

National Research Council (1984) Nutrient requirements of beef cattle. National Academy Press, Washington DC

Naudot C (1977) Camargue et Gardians. Parc Naturel Régional de Camargue, Arles, France

Ngethe JC (1976) Preference and daily intake of 5 East African grasses by zebra. J. Range Mgmt 29: 510–511

Nix HA (1983) The climate of tropical savannas. In Bourliere F (ed) Tropical savannas. Ecosystems of the world 13. Elsevier, Amsterdam pp 37–62

Noy-Meir I (1978) Grazing and production of seasonal pastures. J. Appl. Ecol. 15: 805–835

Oh HK, Jones MB, Longhurst WM (1968) A comparison of rumen microbial inhibition resulting from various essential oils isolated from relatively unpalatable species. Appl. Microbiol. 16(1): 39

Olsen FW, Hansen RM (1977) Food relation of wild free-roaming horses to livestock and big game, Red Desert, Wyoming. J. Range Manage 30(1): 17–20

Olsson N, Kihlen G, Cagell W (1949) Digestibility experiments on horses and evacuation experiments to investigate the time required for the food to pass through the horse's digestive tract. Lantbrukshogskolan, Husdjursforsoksanstalten-Meddelande 36: 5–51

Oosterveld P (1983) Eight years of monitoring of rabbits and vegetation development on abandoned arable fields grazed by ponies. Acta. Zool. Fennica 174: 71–74

Owaga ML (1975) The feeding ecology of wildebeest and zebra in Athi–Kaputei plains. E. Afr. Wildl. J. 13: 375–383

Owaga MLA (1977) Comparison of analysis of stomach contents and faecal samples from zebra. E. Afr. Wildl. J. 15: 217–222

Owen-Smith RN (1985) Niche separation among African ungulates. In Vrba ES (ed) Species and speciation. Transvaal Museum Monograph No. 4, Transvaal Museum, Pretoria, South Africa pp 167–171

Owen-Smith RN (1988) Megaherbivores: the influence of very large body size on ecology. Cambridge University Press, Cambridge, England

Owen-Smith RN, Novellie P (1982) What should a clever ungulate eat? Amer. Naturalist 119: 151–178

Packer C (1985) Dispersal and inbreeding avoidance. Anim. Behav. 33: 676–678

Pader J (1890) La Camargue, son présent, son avenir. Masson, Paris, France

Panthier R, Hanoun C, Oudar J, Beytout D, Corinou B, Joubert L, Guillon J-C,

Mouchet J (1966) Isolement du West Nile virus chez un cheval atteint d'encéphalomyélite. C.R. Acad. Sci. (Paris) 262: 1308–1310

Pennycuick CJ (1979) Energy costs of locomotion and the concept of "foraging radius". In Sinclair ARE, Norton-Griffiths M (eds) Serengeti: dynamics of an ecosystem. University of Chicago Press, Chicago pp 164–184

Penzhorn BL (1982a) Habitat selection by Cape mountain zebras in the Mountain Zebra National Park. S. Afr. J. Wildl. Res. 12: 48–54

Penzhorn BL (1982b) Home range sizes of Cape mountain zebras (*Equus zebra zebra*) in the Mountain Zebra National Park. Koedoe 25: 103–108

Penzhorn BL (1984) A long term study of social organisation and behaviour of Cape mountain zebras (*Equus zebra zebra*). Z. Tierpsychol. 64: 97–146

Peterson R, Mountfort G, Hollom PA D. (1967) Guide des oiseaux d'Europe. Delachaux et Niestlé, Neuchatel, France

Picon B (1977) La Camargue: affectations de l'espace, mutations économiques et rapports sociaux. Centre National de la Recherche Scientifique, Laboratoire d'économie et de sociologie du travail, 13100 Aix-en Provence, France

Picon B (1978) L'espace et le temps en Camargue. Essai d'écologie sociale. Editions Actes Sud, Arles

Pirot J-Y, Chessel D, Tamisier A (1984) Exploitation alimentaire des zones humides de Camargue par cinq espèces de canards de surface en hivernage et en transit: modélisation spatio-temporelle. Rev. Ecol.—Terre Vie 39: 168–192

Poinsot-Balaguer N, Bigot L (1980) Influence of trampling of a horse *manade* in Camargue on the soil fauna and the fauna of canopy. In Soil biology as related to land-use practices. Washington, D.C. pp 358–360

Pollock JI (1977) The ecology and sociology of feeding in *Indri indri*. In Clutton-Brock TH (ed) Primate ecology: studies of feeding and ranging behaviour in lemurs, monkeys and apes. Academic Press, London pp 38–69

Pollock JI (1980) Behavioural ecology and body condition changes in new forest ponies. Royal Society for the Protection of Animals. Scientific Publications No. 6

Poulle M (1817) Etude de la Camargue ou statistique du delta du Rhône. Rapport d'un Ingénieur des Ponts et Chaussées.

Prat H (1954) Vers une classification naturelle des Graminées. Bull. Soc. Bot. France 107: 32–79

Prins RA, Lankhurst A, van Hoven W (1984) Gastrointestinal fermentation in herbivores and the extent of plant cell-wall digestion. In Gilchrist FM C, Mackie RI (eds) Herbivore nutrition in the subtropics and tropics. Science Press, Johannesburg pp 408–434

Pulliam HR, Caraco T (1984) Living in groups: is there an optimal group size? In Krebs JR, Davies NB (eds) Behavioural ecology: an evolutionary approach 2nd edn. Blackwell Scientific Publications, Oxford pp 122–147

Putman RJ (1986) Grazing in temperate ecosystems: large herbivores and the ecology of the new forest. Croom and Helm, London and Sydney

Putman RJ, Edwards PJ, Mann JC E, How RC, Hill SD (1989) Vegetational and faunal changes in an area of heavily grazed woodland following relief of grazing. Biol. Cons. 47: 13–32

Putman RJ, Pratt RM, Ekins JR, Edwards PJ (1987) Food and feeding behaviour of cattle and ponies in the new forest, Hampshire. J Appl. Ecol 24: 369–380

Pyke GH, Pulliam HR, Charnov EL (1977) Optimal foraging: a selective review of theory and tests. Q. Rev. Biol. 52: 137–154

Ralston SL (1984) Controls of feeding in horses. J. Anim. Sci. 59(5): 1354–1361

Ranwell DS (1972) Ecology of salt marshes and sand dunes. Blackwell Scientific Publications, Oxford

Rawes M (1981) Further effects of excluding sheep from high-level grasslands in the north Pennines. J. Ecol. 69(2): 651–669

Rawes M (1983) Changes in two high altitude blanket bogs after cessation of sheep grazing. J. Ecol. 71(1): 219–235

Raymond HL (1978) Contribution a l'étude des Tabanidae (Diptera) de Camargue. Terre Vie 32(2): 291–303

Reid RL, Jung GA (1982) Problems of animal production from temperate pastures. In Hacker JB (ed) Nutritional limits to animal production from pastures. Commonwealth Agricultural Bureau, Farnham Royal, U.K. pp 21–43

Reiner RJ, Urness PJ (1982) Effect of grazing horses managed as manipulators of big game winter range. J. Range Manage. 35(5): 567–571

Reppert JN (1960) Forage preference and grazing habits of cattle at the eastern Colorado Range Station. J. Range Mgmt. 13: 58

Rhoades DF (1983) Herbivore population dynamics and plant chemistry. In Denno RF, McClure MS (eds) Variable plants and herbivores in natural and managed systems. Academic Press, London pp 155–220

Rioux JA, Arnold M (1955) Les Culicidés de Camargue. Terre Vie 9: 244–286

Rioux JA, Corre JJ, Descous S (1968) Application des méthodes phytoécologiques à la détection et l'étude des biotopes larvaires des Diptères hématophages du genre *Leptoconops* (Diptères, Ceratopogonidae). Comple Rendu de l'Académie des Sciences Paris 267: 1219–1222

Risser PG (1969) Competitive relationships among herbaceous grassland plants. Bot. Rev. 35: 251–284

Rittenhouse LR, Johnson DE, Borman MM (1982) A study of food consumption rates and nutrition of horses and cattle. Bureau of Land Management, U.S. Department of Interior, Washington, DC 20240

Robbins CT (1983) Wildlife feeding and nutrition. Academic Press, Orlando, FL

Robinson DW, Slade LM (1974) The current status of knowledge on the nutrition of equines. J. Anim. Sci. 39(6): 1045–1066

Robinson JG (1982) Intrasexual competition and mate choice in primates. Amer. J. Primatol. Suppl. 1: 131–144

Rogers PM (1979) Ecology of the European wild rabbit *Oryctolagus cuniculus* (L.) in the Camargue, southern France. Ph.D. thesis, microfilm, National Library of Canada Ottawa, University of Guelph

Rogers PM (1979) Ecology of the European wild rabbit *Oryctolagus cuniculus* (L.) in the Camargue, southern France. Ph.D. thesis, microfilm, National Library of Canada, Ottawa. University of Guelph, Canada

Rohweder DA, Jorgenson N, Barnes KF (1983) Proposed hay standards based on laboratory analyses for evaluating forage quality. In Smith JA, Hays VW (eds) Proc IVth Int. Grassld Congr. pp 534–538

Romer AS (1966) Vertebrate paleontology. University of Chicago Press, Chicago

Rosenthal GA, Janzen DH (eds) (1979) Herbivores: their interaction with secondary plant metabolites. Academic Press, New York

Roustan C (1807) Les observations sur les chevaux et les haras de Camargue. Mémoire de l'Académie de Marseille, Marseille, France

Rowell TE (1988) Beyond the one-male group. Behaviour 104(3–4): 189–201

Rozin P, Kalat JW (1971) Specific hungers and poison avoidance as adaptive specialisations of learning. Psychological Review 78: 459–486

Rubenstein DA (1981) Behavioural ecology of island feral horses. Equine Vet. J. 13(1): 27–34

Rubenstein DI (1986) Ecology and sociality in horses and zebras. In Rubenstein DI, Wrangham RW (eds) Ecological aspects of social evolution. Princeton University Press, Princeton, NJ pp 282–302

Rubenstein DI, Hohmann ME (1989) Parasites and social behaviour of island feral horses. Oikos 55(3): 312–320

Ruckebusch Y (1972) The relevance of drowsiness in the circadian cycle of farm animals. Anim. Beh. 20(4): 637–643

Ruckebusch Y (1976) Veille—sommeil et environnement chez les équidés. CEREOPA journée d'étude du 5 mars 1976 CEREOPA, 16, rue Claude Bernard, 75231 Paris Cedex 05

Ruckebusch Y, Barbey P, Guillemot P (1970) Les états de sommeil chez le cheval (Equus caballus). Comptes Rendus des Séances de la Société de Biologie. 164(3): 658–665

Ruckebusch Y, Bueno L (1973) Un dispositif simple et autonome d'enregistrement de l'activité alimentaire chez les bovins au paturage. Ann. Rech. vétér. 4: 627–636

Rüger A, Prentice C, Owen M (1986) Results of the IWRB International Waterfowl Census 1967–1983. IWRB special publication No. 6, International Waterfowl and Wetlands Research Bureau. Slimbridge, U.K.

Ryder OA, Epel NC (1978) Chromosome banding studies of the Equidae. Cytogenet. Cell Genet. 20: 323–350

Ryder OA, Sparks RS, Sparks MC, Clegg JB (1979) Hemoglobin polymorphism in Equus przewalskii and E. caballus analysed by isoelectric focussing. Comp. Biochem. Physiol. 62B: 305–308

Salathe T (1985) Nistokolgie des Blasshuhns (Fulica atra) in der Camargue. Ph.D. thesis, University of Basle, Switzerland

Salter RE, Hudson RJ (1978) Habitat utilisation by feral horses in western Alberta. Naturaliste Canadien. 105(5): 309–321

Salter RE, Hudson RJ (1979) Feeding ecology of feral horses in Western Alberta. J. Range Manage. 32(3): 221–225

Salter RE, Hudson RJ (1980) Range relationships of feral horses with wild ungulates and cattle in Western Alberta. J. Range Manage. 33(4): 266–270

Schaeffer B (1948) The origin of a mammalian ordinal character. Evolution 2: 164–175

Schaller GB (1972) The Serengeti lion. University of Chicago Press, Chicago

Schein MW, Fohrman MH (1955) Social dominance relationships in a herd of dairy cattle. Br. J. Anim. Behav. 3: 45–55

Seal US, Foose T, Lacy RC, Zimmermann W, Ryder O, Princee F (1990) Przewalski's horse (Equus przewalskii). Global conservation plan—draft. Captive Breeding Specialist Group, Species Survival Commission, IUCN-World Conservation Union, Switzerland

Seal US, Plotka ED (1983) Age-specific pregnancy rates in feral horses. J. Wildl. Mgmt. 47: 422–429

Senft RL, Rittenhouse LR (1985) A model of thermal acclimation in cattle. J. Anim. Sci. 61(2): 297–305

Sereni J-L (1977) Recherches sur l'évolution de la structure sociale et du comportement alimentaire d'un groupe de chevaux camarguais. Thèse Doctorat, Université d'Aix-Marseille II, France

Sereni J-L, Bouissou MF (1978) Mise en évidence des relations de dominance-subordination chez le cheval par la méthode de compétition alimentaire par paire. Biology of Behaviour 3: 87–93

Short RV (1975) The evolution of the horse. J. Reprod. Fert., Suppl. 23: 1–6

Sibly RM, McFarland DJ (1976) On the fitness of behavior sequences. Amer. Naturalist 110: 601–617

Siegel S (1956) Nonparametric statistics for the behavioural sciences. McGraw-Hill, New York

Siegfried WR (1978) Habitat and modern range expansion of the Cattle egret. In

Sprunt AI V, Ogden JC, Winckler S (eds) Wading birds. National Audubon Society, New York

Silk JB (1987) Social behaviour in evolutionary perspective. In Smuts B, Cheyney D, Seyfarth RM, Wrangham RW, Struhsaker TT (eds) Primate societies. University of Chicago Press, Chicago pp 318–329

Simpson GG (1951) Horses. Oxford University Press, Oxford, England

Sinclair ARE (1975) Resource limitation of trophic levels in tropical grassland ecosystems. J. Anim. Ecol. 44: 497–520

Sinclair ARE (1979) The eruption of the ruminants. In Sinclair ARE, Norton-Griffiths M (eds) Serengeti: dynamics of an ecosystem. University of Chicago Press, Chicago pp 82–103

Sinclair ARE, Dublin H, Borner M (1985) Population regulation of Serengeti wildebeest: a test of the food hypothesis. Oecologia 65: 266–268

Sinclair ARE, Gwynne MD (1972) Food selection and competition in the East African buffalo (*Syncerus caffer* Sparrman). E. Afr. Wildl. J. 10: 77–89

Sinclair ARE, Norton-Griffiths M (1982) Does competition or facilitation regulate ungulate populations in the Serengeti? a test of hypotheses. Oecologia 53: 364–369

Skelton ST (1978) Seasonal variations and feeding selectivity in the diets of horses (*Equus caballus*) of the Camargue. M.Sc. thesis, Texas A & M University College Station

Skogland T (1984) Wild reindeer foraging-niche organization. Holarctic Ecology 7: 345–379

Slade LM, Hintz HF (1969) Comparison of digestion in horses, ponies, rabbits and guineapigs. J. Anim. Sci. 28: 842–847

Smith CJ (1980) Ecology of the English chalk. Academic Press, London

Smith LM, Kadlec JA (1985) Fire and herbivory in a Great Salt Lake marsh. Ecology 66(1): 259–265

Smuts GL (1975a) Home range sizes for Burchell's zebra, *Equus burchelli antiquorum* from the Kruger National Park. Koedoe 18: 136–149

Smuts GL (1975b) Pre-natal and post-natal growth phenomena of Burchell's zebra, *Equus burchelli antiquorum.* Koedoe 18: 69–102

Smuts GL (1976a) Population characteristics of Burchell's zebra (*Equus burchelli antiquorum*, H. Smith, 1841) in the Kruger National Park. So. Afr. J. Wildl. Res. 6: 99–112

Smuts GL (1976b) Reproduction in the zebra mare. Koedoe 19: 89–132

Snaydon RW (1981) The ecology of grazed pastures. In Morley FH W. (ed) Grazing animals. World Animal Science B1 Elsevier, Amsterdam pp 13–31

Sokal RD, Rohlf FJ (1972) Biometry. Freeman, San Francisco, CA

Sparks DR, Malechek JC (1968) Estimating percentage dry weights in diets using a microscopic technique. J. Range Manage. 21(4): 264–265

Spinage CA (1968) A quantitative study of the daily activity of Uganda defassa waterbuck. E. Afr. Wildl. J. 6: 89–93

Stanley-Price MR (1978) The nutritional ecology of Coke's hartebeest (*Alcelaphus buselaphus cokei*) in Kenya. J. Appl. Ecol. 15: 33–49

Stanley Price M (1989) Animal re-introductions: The Arabian oryx in Oman. Cambridge University Press, Cambridge, England

Stewart DRM (1965) The epidermal characters of grasses with special reference to East African plains species. Bot. Jb. 84(1): 63–116

Stewart DRM (1967) Analysis of plant epidermis in faeces: a technique for studying the food preferences of grazing herbivores. J. Appl. Ecol. 4: 83–111

Stewart DRM, Stewart J (1970) Food preference data by faecal analysis for African plains ungulates. Zoologica Africana 15(1): 115–129

Storr GM (1961) Microscopic analyses of faeces, a technique for ascertaining the diet of herbivorous mammals. Austral. J. Biol. Sci. 14: 157–164

Svejcar TJ (1989) Animal performance and diet quality as influenced by burning on tallgrass prairie. J. Range Manage. 42(1): 11–15

Tamisier A (1971) Régime alimentaire des sarcelles d'hiver, *Anas crecca* L., en Camargue. Alauda 39: 261–311

Tamisier A (1972) Etho-écologie des sarcelles d'hiver Anas. c. crecca L. Pendant leur hivernage en Camargue. Thèse d'etat, Univ de Montpellier, Montpellier, France

Tashiro H, Schwardt HH (1953) Biological studies of horse flies in New York. J. Econ. Ent., 46: 813–822

Taylor DM (1986) Effects of cattle grazing on passerine birds nesting in riparian habitat. J. Range Manage. 39(3): 254–258

Thomas G, Allen DA, Grose MP B (1981) The demography and flora of the Ouse Washes. Biol. Cons. 21: 197–229

Thornthwaite CW (1948) An approach toward a rational classification of climate. Geogr. Rev. 38: 55–94

Thornton RF, Minson DJ (1972) The relationship between voluntary intake and mean apparent retention time in the rumen. Aust. J. Agric. Res. 23: 871–877

Thorsteinson AJ, Bracken GK, Hanec W (1965) The orientation behaviour of horseflies and deerflies (Tabanidae Diptera). The use of traps in the study of orientation of tabanids in the field. Ent. Exp. and Appl. 8: 189–192

Toussaint H (1874) Le cheval dans la station préhistorique de Solutre. Recueil de médécine vétérinaire, 6ème série, tome 1: 380–392 and 467–474

Treisman M (1975) Predation and the evolution of gregariousness: I. Models for concealment and evasion. Anim. Behav. 23: 779–800

Tribe DE (1950) Influence of pregnancy and social facilitation on the behaviour of grazing sheep. Nature 166: 174

Tschanz B (1979) Sozialverhalten beim Camarguepferd. Dokumentierverhalten bei Hengsten (Freilandaufanhmen). Institut für Wissenschaftliche Film Gottingen Sektion Biologie Serie 12, No. 12 Film D 1284

Turner JW Jr., Perkins A, Kirkpatrick JF (1981) Elimination marking behavior in feral horses. Canadien J. Zool. 59(8): 1561–1566

Turner MG (1987) Effects of grazing by feral horses, clipping, trampling and burning on a Georgia saltmarsh. Estuaries 10(1): 54–61

Tyler SJ (1972) The behaviour and social organisation of the New Forest ponies. Anim. Behav. Monogr. 5: 85–196

Ungar ED, Noy-Meir I (1988) Herbage intake in relation to availability and sward structure: grazing processes and optimal foraging. J. Appl. Ecol. 25: 1045–1062

Van Deursen EJM, Drost HJ (1990) Defoliation and treading by cattle of reed *Phragmites australis*. J. Appl. Ecol. 27: 284–297

Van Dyne GM, Torrell DT (1964) Development and use of the oesophageal fistula, a review. J. Range Manage. 17: 7

Van Niekerk CH, Van Heerden JS (1972) Nutrition and ovarian activity of mares early in the breeding season. J. S. Afr. Vet. Med. Ass. 43: 351–360

Van Soest PJ (1965) Symposium on factors influencing the voluntary intake of herbage by ruminants: voluntary intake in relation to chemical composition and digestibility. J. Anim. Sci. 24: 834–843

Van Soest PJ (1982) Nutritional ecology of the ruminants. O & B Books, Corvallis, OR

Van Twyver H, Allison T (1969) EEG study of the shrew (*Blarina brevicauda*): a preliminary report. Psychophysiology 29: 231

Van Wieren, SE (1990) The management of populations of large mammals. In IF Spellerberg, FB Goldsmith and MG Morris (eds) The Scientific Management of

Temperate Communities for Conversation. Blackwell Scientific Publications, Oxford, England, pp. 103–127

Vander Noot GW, Gilbreath EB (1970) Comparative digestibility of components of forages by geldings and steers. J. Anim. Sci. 31: 351–355

Vander Noot GW, Symons LD, Lydman RK, Fonnesbeck PV (1967) Rate of passage of various feedstuffs through the digestive tract of horses. J. Anim. Sci. 26: 1309–1311

Vander Noot GW, Trout JR (1971) Prediction of digestible components of forages by equines. J. Anim. Sci. 33: 38–41

Vandewalle P (1989) Le cycle reproducteur du Lapin de garenne (*Oryctolagus cuniculus*) en Camargue: influence des facteurs environnementaux. Gibier Faune Sauvage 6: 1–25

Vavra M, Sneva F (1978) Seasonal diets of five ungulates grazing the cold desert biome. In Proceedings of the First International Rangelands Congress pp 435–437

Vlassis G (1978) Le cheval Camargue à travers le temps. Courrier du Parc Naturel Régional de Camargue No. 14: 8–10

von Goldschmidt-Rothschild B, Tschanz B (1978) Sozial Organisation und Verhalten einer Jungtierherde beim Camargue-Pferd. Z. Tierpsychol. 46: 372–400

Walker EP (1968) Mammals of the world. Johns Hopkins Press, Baltimore MD

Watt AS (1962) The effect of excluding rabbits from grassland (*Xerobrometum*) in Breckland 1936–60. J. Ecol. 50: 181–198

Webb SD (1977) A history of savanna vertebrates in the New World. I. Ann. Rev. Ecol. Syst. 8: 355–380

Wells SM, von Goldschmidt-Rothschild B (1979) Social behaviour and relationships in a herd of Camargue horses. Z. Tierpsychol. 49: 363–380

Welsh DA (1975) Population, behavioural and grazing ecology of the horses of Sable Island, Nova Scotia. M. Sc. thesis, Dalhousie University, Canada

Western D (1975) Water availability and its influence on the structure and dynamics of a savannal large mammal community. E. Afr. Wildl. J. 13: 265–286

Whyte RJ, Cain BW (1981) Wildlife habitat on grazed or ungrazed small pond shorelines in South Texas. J. Range Manage. 34(1): 64–68

Wittenberger JF, Hunt JL Jr (1985) The adaptive significance of coloniality in birds. In Farner DS, King JR, Parkes KC (eds) Avian biology. VIII Academic Press, New York pp 1–78

Wolfe ML (1979) Population ecology of the Kulan. In Proceedings of Symposium on the Ecology and Behavior of Wild and Feral Equids. Sept. 6–8 University of Wyoming, Laramie pp 205–218

Wolfe ML (1982) Alternative population limitation strategies for feral horses. In Peek JD, Dalke PD (eds) Proc. Wildlife–livestock relationships symposium: forest, wildlife and range experimental station. University of Idaho, Moscow, Idaho pp 394–408

Wolfe ML, Ellis LC, MacMullen R (1989) Reproductive rates of feral horses and burros. J. Wildl. Manage. 53(4): 916–924

Woodburne MO, MacFadden BJ (1982) A reappraisal of the systematics, biogeography, and evolution of fossil horses. Paleobiology 8(4): 315–327

Woodward SL (1979) The social system of feral asses (*Equus asinus*). Z. Tierpsychol. 49: 304–316

Wrangham RW (1980) An ecological model of female-bonded primate groups. Behaviour 75: 262–300

Index

Ecological Studies